D0931644

Smart Card Manufacturing

Smart Card Manufacturing

A Practical Guide

Yahya Haghiri
Thomas Tarantino
Giesecke and Devrient GmbH, Munich, Germany

JOHN WILEY & SONS, LTD

Copyright 2002 © John Wiley & Sons, Ltd
Baffins Lane, Chichester
West Sussex, PO19 1UD, England

National 01243 779777
International (+44) 1243 779777
e-mail (for orders and customer service enquiries): cs-books@wiley.co.uk

Visit our Home Page on http://www.wileyeurope.com or http://www.wiley.com

Other Wiley Editorial Offices

John Wiley & Sons, Inc., 605 Third Avenue,
New York, NY 10158-0012, USA

Wiley-VCH Verlag GmbH
Pappelallee 3, D-69469 Weinheim, Germany

Jacaranda Wiley Ltd, 33 Park Road, Milton,
Queensland 4064, Australia

John Wiley & Sons (Canada) Ltd, 22 Worcester Road
Rexdale, Ontario, M9W 1L1, Canada

John Wiley & Sons (Asia) Pte Ltd, 2 Clementi Loop #02-01,
Jin Xing Distripark, Singapore 129809

Library of Congress Cataloging-in-Publication Data
Haghiri, Yahya
Smart card manufacturing : a practical guide / Yahya Haghiri, Thomas Tarantino
    p. cm.
Translation of Vom Plastik zur Chipkarte
Includes bibliographical refernces and index
ISBN 0 471 49767 3
    1.   Smart cards I. Tarantino, Thomas II. Title

TK7895.S62 H34 2001
621.39'16 – dc21

                                                    2001057382

British Library Cataloguing in Publication Data

A catalogue record for this book is available from the British Library

ISBN 047149767 3

Contents

Preface

Inexorably the smart card penetrates into new areas of our daily life. The calling card was the first widespread application of the smart card in Europe. The latest introduction of the patient insuring card and access card for mobile phones means the smart card has become a natural constituent of modern life.

Initial doubts raised against the smart card, e.g. that they reduce the human citizen to a transparent person, are discussed today leading to the demand for personal data security. The smart card is considered as a suitable means, which can serve the protection of the individual in today's very complex systems.

The application possibilities of the smart card are widespread. They cover a large variety of applications, e.g. conditional access to security areas, pay-TV, PC access and authorization for transactions and protection of data communication in public (Internet) and private (Intranet) networks.

The varied application of the smart card in increasing areas requires its constant advancement. This is especially the case for increase of intelligence and performance of the smart card for better protection of the owner's data in public communication networks.

Today, there are several publications which describe the electronic functionality of the smart card and smart card system. However, to our knowledge there is no literature available which describes the way smart cards are produced; which material, machines concepts and technologies are used.

This text attempts to close this gap by leading the reader step by step through the development of smart cards out of sand and plastic.

We wish to point out that most solutions represented in this text are protected by patents. For commercial use of these solutions the legal situation should be considered.

We wish our readers fun in the world of the smart card.

During translation and updating from the German version of this text Yahya Haghiri had a tragic accident. This book is dedicated in honour of a great man who inspired the development of smart cards over many years, and to a friend.

Thomas Tarantino
Laufen, December 2001

1 Function and structure of a smart card

In this chapter we want to explain the principle of smart cards using the examples of a few selected basic smart cards and their body constructions. Before that, however, explanations to some definitions are required.

There are many different names for the smart card, for example IC-card, microprocessor card, electronic card, etc. We would like to use the term 'smart card' for a card which contains a semiconductor device (chip) and a data link for data communication between the smart card and smart card reader.

This connection can be established via outside contact areas. These smart cards are called contact smart cards. In addition, the connection also can take place without visible contacts so-called 'contactless smart cards'.

A contactless data connection can be realized inductively over an antenna, capacitively with a condenser, optically with an opto coupler or by other physical effects.

Also several connecting types are possible at the same time or optionally (for example with contact and inductively).

1.1 Smart cards with contacts

Substantial components of a contact smart card are the card body, the chip and the outside contacts. On the basis of a common version of the smart card the structure, in principle, can be described. Further explanations concerning connection techniques and manufacturing techniques are represented in the following chapters.

1.1.1 Structure of a smart card with contacts

Figure 1.1 schematically shows a common version of a contact smart card. A cavity is produced in the card body, for example, by milling. Into this cavity a chip module is bonded with an adhesive. The thickness of the card is drawn disproportionately bigger for a better overall view.

Figure 1.2 shows the closer surrounding of the chip module enlarged. The chip module contains a chip, which is connected to the outside contacts through thin gold wires (bond wire).

The contacts are defined by international standards in their number, size and position, hence the function of each smart card in each smart card reader is guaranteed over the world.

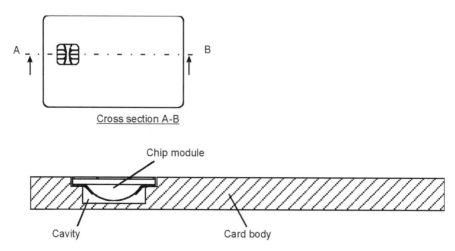

Cross section A-B

Figure 1.1 Schematic structure of a contact smart card

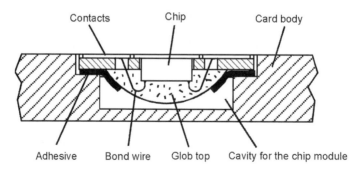

Figure 1.2 Enlarged cross section of the chip module area of a smart card

Figure 1.3 shows the most important dimensions of a smart card. In the standards, eight contacts are determined for the smart card and indicated from top left to left down by 1—4 as well as from top right to right down by 5—8. The functions of contacts 1—8 are defined as follows:

1	VCC	Operating voltage
2	RST	Reset
3	CLK	Clock
4	—	n/a
5	GND	Ground
6	V_{PP}	Programming voltage
7	I/O	Data input/output
8	—	n/a

For contacts 4 and 8 so far there are no functions determined in the standards. Therefore, if possible, chip modules with only six contacts are implemented in most cases. Usually the focal point of the upper six contacts is measured as shown in Figure 1.3. For a better survey, the dimensions in this figure are unlike those described in the standards. In the standards, every single contact is measured with a minimum and maximum distance between the reference points at the edge of the smart card.

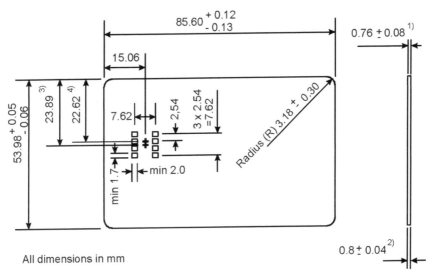

All dimensions in mm

Figure 1.3 Overview of the most important dimensions of the smart card
(1) Thickness of the card according to the standard
(2) Common thickness tolerances of the card body for smart cards
(3) Focal point for chip modules with eight contacts
(4) Focal point for chip modules with six contacts

Further with most chips the programming voltage is produced chip-internal, so that contact six mostly is not occupied and remains without function.

The thickness of the smart card according to the standard is 0.76 ± 0.08 mm. For the production of smart cards with chip modules, however, card bodies are used in the upper range of tolerance. In Figure 1.3 these tolerances are indicated as 0.8 ± 0.04 mm.

1.1.2   Standards

The most important international standards for smart cards are listed in the following. ISO (International Organization for Standardization) and IEC (International Electrotechnical Commission) are international standardization organizations. They cooperate in common fields and coordinate the work of national standard committees like the German DIN or the French AFNOR.

ISO/IEC 7810	Identification cards — Physical characteristics
ISO/IEC 7811-1/2/3/4/5/6	Identification cards — Recording technique
ISO/IEC 7811-1	Part 1: Embossing
ISO/IEC 7811-2	Part 2: Magnetic stripe
ISO/IEC 7811-3	Part 3: Location of embossed characters on ID-1 cards
ISO/IEC 7811-4	Part 4: Location of read-only magnetic tracks — Tracks 1 and 2
ISO/IEC 7811-5	Part 5: Location of read-write magnetic track — Track 3
ISO/IEC 7811-6	Part 6: Magnetic stripe — High coercivity
ISO/IEC 7812-1/2	Identification cards — Identification of issuers
ISO/IEC 7812-1	Part 1: Numbering system
ISO/IEC 7812-2	Part 2: Application and registration procedures
ISO/IEC 7813	Identification cards — Financial transactions cards
ISO 7816-1/2/3/4/5/6/7/8/9/10	Identification cards — Integrated circuit(s) cards with contacts
ISO 7816-1	Part 1: Physical characteristics
ISO 7816-2	Part 2: Dimensions and location of the contacts
ISO 7816-3	Part 3: Electronic signals and transmission protocols
ISO 7816-4	Part 4: Interindustry commands for interchange
ISO 7816-5	Part 5: Numbering system and registration procedure for application identifiers
ISO 7816-6	Part 6: Interindustry data elements
ISO 7816-7	Part 7: Interindustry commands for structured card query language (SCQL)
ISO 7816-8	Part 8: Security-related interindustry commands
ISO 7816-9	Part 9: Additional interindustry commands and security attributes
ISO 7816-10	Part 10: Electronic signals and answer to reset for synchronous cards
ISO/IEC 10373-1/2/5	Identification cards — Test methods
ISO/IEC 10373-1	Part 1: General characteristics tests
ISO/IEC 10373-2	Part 2: Cards with magnetic stripes
ISO/IEC 10373-5	Part 5: Optical memory cards
ISO/IEC 10536-1/2/3	Identification cards — Contactless integrated circuit(s) cards — Close-coupled cards
ISO/IEC 10536-1	Part 1: Physical characteristics
ISO/IEC 10536-2	Part 2: Dimensions and location of coupling areas
ISO/IEC 10536-3	Part 3: Electronic signals and reset procedures
ISO/IEC 14443	Identification cards — Contactless integrated circuit(s) cards — Proximity cards
ISO/IEC 14443-1	Part 1: Physical characteristics
ISO/IEC 15693-1/2	Identification cards — Contactless integrated circuit(s) cards — Vicinity cards

ISO/IEC 15693-1	Part 1: Physical characteristics
ISO/IEC 15693-2	Part 2: Air interface and initialization

Beyond these general standards there are also standards for special areas of application as, for example, bank cards or telecommunications cards; also separate specifications for certain applications such as credit cards, Eurocheque card (ec-card) and calling cards, which are distributed in each case from the card organizations responsible.

1.1.3 Transmission protocols

The transmission protocols for smart cards are defined in ISO/IEC 7816-3. This standard determines the electrical functions of the contacts, defines the type of contacting and activation of the contacts in the reader, and describes the response of the card to resetting (answer to resets) during synchronous or asynchronous transfer.

Further in ISO/IEC 7816-3 the different types of the transmission protocols T are defined. The different transmission protocols developed historically from the different developments of this area.

The protocols numbered in the standards range from 0 to 15 (16 numbers = 8 bits = 1 Byte); the numbering relates to the fact that the smart card can signal the type of transfer in a simple manner to the reader.

$T = 0$	Asynchronous half duplex character transmission protocol
$T = 1$	Asynchronous half duplex block transmission protocol
$T = 2$ and $T = 3$	Reserved for future full duplex operations
$T = 4$	Reserved for an enhanced asynchronous half duplex character transmission protocol
$T = 5... T = 13$	Reserved for future uses
$T = 14$	Reserved for protocols standardized by ISO
$T = 15$	Reserved for future extension

The transmission protocols are described in paragraph 8 of the standard ISO/IEC 7816-3.

1.1.4 Different types of smart cards

Depending upon the application different chips are used in smart cards. Chips are distinctive according to the type of memory, the CPU and their interface to the reading devices. In the following the different types of smart cards are listed.

Memory smart cards

Memory smart cards contain memory chips. They usually consist of non-volatile memory cells, a special logic and an interface for input and output of the data.

The non-volatile memory is called EEPROM or E^2PROM.

EEPROM = ELECTRICALLY ERASABLE PROGRAMMABLE READ-ONLY
          MEMORY

These memory cells can be programmed electrically and keep their memory contents after power-off of the current. Forerunners of the smart cards with EEPROM were smart cards with EPROM.

EPROM = ELECTRICALLY PROGRAMMABLE READ-ONLY MEMORY

These memories can be electrically coded. However, the once-coded memory can be reprogrammed afterwards only by ultraviolet light.

The memory chips used in the smart cards are usually quite small (a size of e.g. 1 mm^2) and therefore economical. However the microprocessors are much larger because of their higher functionality (normally e.g. 25 mm^2).

Microprocessor smart cards

Microprocessor smart cards contain microprocessor chips. The microprocessor chip is structured like a computer with different memory types and a CPU.

CPU = CENTRAL PROCESSING UNIT

The CPU of the microprocessor can calculate the stored data following programmed rules and arrange and control the data. These microprocessors offer a larger functionality and a higher security against manipulation than memory smart cards.

Besides the EEPROM mentioned, further storage types are accommodated such as ROM and RAM in the microprocessors.

ROM = READ ONLY MEMORY

ROMs have a constant content, which cannot be changed. Their content is given by production masks in the production process of the chips. A modification of ROM can only take place via a new edition of the chip with modified masks.

RAM = RANDOM ACCESS MEMORY

RAM is volatile memory, which can be reprogrammed electrically, and functions during operation as a working memory.

Crypto controller

Cryptoprocessors are microprocessors which include additional possibilities for encoding and decoding data.

A cryptoprocessor is able to execute the calculation of keys in the required length cryptographic procedures in very short time. The cryptoprocessor is usually integrated additional to the CPU as a coprocessor.

Newest memory for smart cards is offered by FERAM technology developed by Fujitsu and Ramtron.

FERAM = FERRO ELECTRICAL RAM

FERAM memory can be used as working memory just like a non-volatile memory. The advantage of FERAM memory technology compared to EEPROM is that the writing of data into the memory can be executed much faster and much more frequently. Additionally they have smaller power absorption than EEPROM, which is advantageous for the contactless smart card.

1.1.5 Applications

In 1998 worldwide approximately 1.1 billion smart cards were manufactured. This number was expected to rise to 1.5 billion in 2000.

The rate of growth of turnover with chips for smart cards is estimated at 36 % per year. As the price of the chip makes up 50 to 70 % of the smart card price, a rate of growth of approximately 15 % can be assumed, whereby the in sales do not correlate necessarily with the increased number of smart cards, but with the in connection higher functionality of the smart cards. For worldwide produced smart cards for different applications an average growth rate of 36 % per year is estimated to 2003, where as smart cards for information technology with 142 % growth rates are cited — far over average.

In 1987 almost exclusively memory chip cards were predominant in use as calling cards, today microprocessor cards achieve about 75 % of the worldwide turnover. This tendency will continue to strengthen in the future, whereby a part of these microprocessor cards will have an additional contactless interface and also have cryptoprocessors.

In the following some applications of smart cards are presented briefly.

Calling cards

The smart card as telephone value card is usually a memory chip card with a safety logic. The first calling cards, which were used in France, worked with EPROM memories which were cancelled successively during use in the telephone box. Today's systems use EEPROM memory, which can be rewritten several times, but because of appropriate logical barriers the determined number of units offered decreases during a phone call and cannot be reloaded again.

Card telephone boxes are replacing public telephone coin boxes. For the operator, substantial advantages of the card telephone compared to the telephone coin box are:

- Telephone devices become simpler; no complex coin slots required
- Collection of the coins is omitted
- Easy change of fees
- Telephone charges are paid in advance with purchase of the card
- Reduced violence against telephone boxes

Also for the user calling cards bring many advantages. The annoying refill of coins during a telephone call and inadvertent overpayment or telephone call interruptions and the search for change almost belong to the past.

While more and more countries change over to card telephones, whereby total number of calling cards rises worldwide, the tendency is stagnating or even declining with telephone value cards in pioneer countries like France and Germany. The reason is due on the one hand to proliferation of mobile telephones whereby the requirement for public telephones decreases. On the other hand, more and more telephone customers are using the possibility of making a telephone call with other card systems as for example with the reloadable smart card.

Health care card (patient card)

The German patient insuring card (see Figure 1.4) only contains patient-relevant data for the physician account, justified for the physician attendance, along with a memory chip. In some other countries microprocessors are used for this, which fulfil various health service functions beyond the physician account and contain additional information such as disease history, blood group, allergies and other emergency information.

Figure 1.4 Sample of a German patient insuring card (source: Giesecke & Devrient)

GSM smart card

GSM (Global System For Mobile Communication) is the name for a spread-standardized mobile phone network in Europe, some Asiatic and African, countries and in some parts of the Unites States of America. GSM standards determine a 'subscriber identity module' (SIM) for the identification of the user in the network in the form of a microprocessor card. The SIM card contains the necessary security functions for user identification and various additional functionalities for the mobile phone.

As alternative to the smart card in ISO standard GSM gives rise to a small smart card, which is termed 'plug-in SIM' for small mobile telephones. Figure 1.5 shows the dimensions of the 'plug-in SIM' in comparison to the ISO smart card (ID1) represented.

In practice, the plug-in SIM is punched out from an ISO smart card ID1. There are also versions which are pre-punched so that both SIM and plug-in SIM can be used.

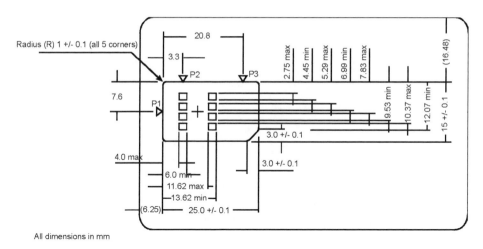

All dimensions in mm

Figure 1.5 Dimensions of the plug-in SIM in an ID1 card. P1, P2 and P3 are points of reference for the position of the contacts

Electronic purse

The electronic purse — recently implemented with the Eurocheque card in Germany — offers the owner the possibility to pay certain amounts off-line with the smart card, i.e. without account query and without input of a PIN (personal identification number) number. To credit money on the smart card the card holder has to go to an ATM (automated teller machine) and insert the smart card into the card slot and then can transfer a certain amount from his/her bank account to the smart card. This amount can be paid in any fraction at the dealer terminals.

JAVA smart card

JAVA is a higher programming language which was developed by JAVA SOFT, a daughter of SUN Microsystems. This language is particularly suitable for programming processes in communications networks such as the Internet. Using this language, a so-called 'virtual machine' can be programmed which can control safe loading of programs from a network. The use of JAVA in the smart card opens the possibility of downloading new applications into the smart card after the smart card was issued to the card holder. Thus JAVA creates an open, future-orientated multi-functional platform smart card.

Figure 1.6 shows a JAVA smart card.

Cryptoprocessor smart cards

Cryptoprocessor cards are smart cards with a cryptoprocessor which can be used for encoding and decoding data by use of cryptographic algorithms.

The principle of cryptology consists of the fact that the message in transmission place **A**, which can be transmitted from place **A** to place **B**, is transformed in a way that it is not recognizable, and in receipt place **B** transformed back to the origin form.

Figure 1.6 Example of a JAVA smart card (source: Giesecke & Devrient)

Figure 1.7 represents a simplified form of the encoded data communication.

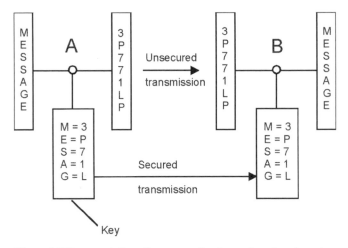

Figure 1.7 Representation of easy encoding by exchanging characters

The encoding procedure in this case is the fact that the alphanumeric characters are exchanged unordered. The code is in this case a list representing allocation of the alphanumeric characters. The security of the procedure depends primarily on whether the code is securely transferred and also stored securely. Further, it is most important that the procedure remains secret; as soon as it is apparent that the exchanging of characters has taken place and, if additionally, the language of the message is known, then statistical rules can be used which facilitate the search. For example, in the German language combination of the two letters **CH** enables a faster search. The accumulation of occurrence of each letter in different languages is statistically stored and wellknown. With very long messages these statistics are a first assistance for breaking the encoding. By use of efficient computers the encoding may be found in an acceptable time.

The security of the encoding is measured in the time which an attack would need to unencrypt the message under usage of high-performance computer systems and systematic trying. If the necessary computer performance for encryption is larger than all overall available computer performances, then this encoding procedure is considered as secure enough. If by technological progress the available computer performance rises precipitously or new procedures for decoding are developed by advancement of cryptology, it is no longer safe enough.

Besides the secure encoding procedure it is also important to keep secret the elements which are used for encoding. But the greater the number of people involved in the procedure the greater the possibility, that secret information gets published. Therefore for modern systems, in which a multiplicity of people participate, encoding procedures were developed with which each user has his/her own secret code for which only the data for this special user can be decoded. The public key procedure belongs to this category. In the public key procedure for each user a secret key and a public key is generated. The secret key is transmitted to the user using a safe way, and the public key is put in a directory for the usage of all users.

Figure 1.8 represents this procedure schematically.

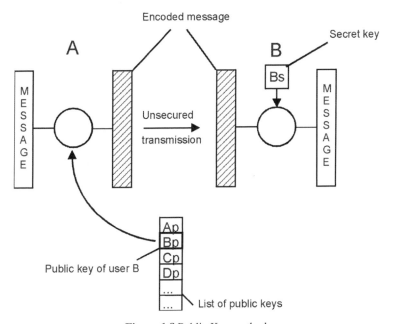

Figure 1.8 Public Key method

User **A**, would like to transmit an encoded message to user **B**, and encodes his message with the public key **Bp** of user **B**. The algorithm for this encoding is laid out in a way that user **B** can decode the encoded message with his own secret code **Bs**. The algorithm is not only wellknown in this way, but also standardized. Security exists not in the secrecy of the procedure and the algorithm, but on the one hand in the secrecy of the key and on the other hand in the security of the algorithm as well as in the length of the key. The security of the algorithm is constantly checked by expert committees. They make suggestions on the necessary length of the key, therefore a promising 'hacker' attack requires an arithmetic performance which cannot be supplied by a computer at the moment.

The cryptoprocessor smart card offers the possibility of storing the key within the smart card and of executing the calculations at least partly in the smart card. It is not necessary for the user to type the secret key into the system, risk being watched and have the key stolen.

Figure 1.9 shows two types of cryptoprocessor smart cards. The STARCOS operating system of Giesecke & Devrient is particularly developed for smart cards on whose base cryptographic applications for example a digital signature were developed.

With the application of a digital signature the encoding is used in order to secure a document after signing, which is available as a file, and to guarantee the authenticity of the digital signature and the authenticity of the document.

Figure 1.9 Two examples of cryptoprocessor smart cards (source: Giesecke & Devrient)

Further applications are home banking via Internet, shopping in the Internet (e.g. software, audio, video, books and tickets), or communication of confidential data (e.g. companies, authorities, physicians and lawyers).

1.2   Contactless smart card

Figure 1.10 schematically shows the principle of the contactless smart card system with a reader and a contactless smart card.

The contactless smart card consists of a transponder (antenna) which is located inside the card. The transponder consists of a chip which is attached to an antenna.

The contactless smart card communicates with the reader via electromagnetic waves. The smart card functions like a transmit and receive device. In the reader is also inserted a transmitter/receiver station.

The electromagnetic waves sent by the reader produce an oscillating electromagnetic field. This field induces an oscillating electrical voltage in the antenna of the smart card.

With a basic amount of voltage the chip is supplied with current, and the oscillations of the voltage are caught as a signal and transferred internally into data in the chip. These data are processed within the chip and transferred as modifications of the magnetic field, which are caught again by the reader and transferred into data.

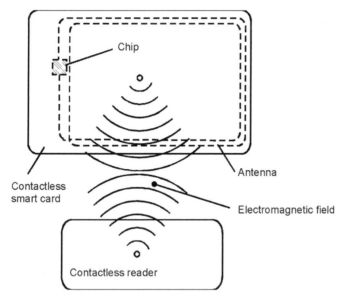

Figure 1.10 Function of the contactless smart card system

Beyond the shown example there are contactless systems on the market which have one antenna for energy transmission and one antenna for signal (data) transmission.

However, systems with only one antenna, which transmits both signals and the energy for signal processing in the contactless smart card, have become generally accepted. This technology is used in the Mifare® systems from Philips, LEGIC® from the Kaba Security Lockin System and the contactless system GO Card® from CUBIC.

1.2.1 Structure of a contactless smart card

As already described, at least one chip and an antenna in the smart card belong to the contactless smart card. Today many innovative ideas for realization of the antenna and contacting of the chip with the antenna have become wellknown (for this see also Section 7.2). The following example refers to a present use realization of the contactless smart card.

Figure 1.11 shows the structure of a contactless smart card. The chip is built into a chip module in a preceding work step and the antenna is applied on a plastic foil on the card. The chip module is contacted to the antenna on the inlay. Subsequently, one balance foil and two overlay foils are mounted in addition. For accommodation of the chip module in the inlay and in the balance foils a free punched area with the size of the encapsulation and contacts of the chip module is created. In order to avoid the chip module causing a short circuit of the turns of the antenna, the antenna within the chip module area is covered with an insulation material. The electrical connection of the chip module contacts to the antenna can be made, for example, by partial dosage of a conductive adhesive.

The so-arranged package of foils is laminated, i.e. clamped between two metal plates, and heated up to the softening point of the foil material and afterwards cooled down in the pressed state. The result is a sheet, where the contactless smart card can be punched out.

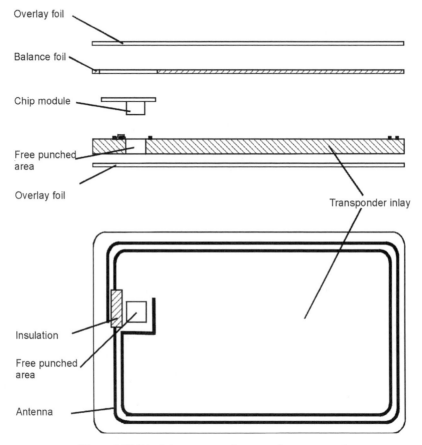

Figure 1.11 Principle structure of a contactless smart card

Figure 1.12 shows magnified the principle of a contactless chip module. The chip is glued on a punched or etched metal lead frame. The electrical contacts of the chip (pads) are connected electrically (bonded) to the two contact plates of the lead frame. This connection is made via very fine gold bond wires which are welded at the chip contacts and the contacts of the lead frame by a special micro-welding method (bond technique). The chip, the bond wires and the surrounding of the contacts on the lead frame are covered in each case with a resin and form a chip module.

A possible technique for realization of the antenna is screen printing, whereby instead of colour a conductive paste with silver powder and binding agent is printed.

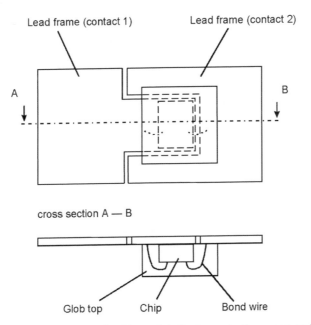

Figure 1.12 Principle of a chip module for the contactless smart card

1.2.2   Standards

For the contactless smart card the following standards are valid:

ISO/IEC 10536-1/2/3   Identification cards — Contactless integrated circuit(s)
                           cards — Close-coupled cards
   ISO/IEC 10536-1   Part 1:   Physical characteristics
   ISO/IEC 10536-2   Part 2:   Dimensions and location of coupling areas
   ISO/IEC 10536-3   Part 3:   Electronic signals and reset procedures

ISO/IEC 14443-1   Identification cards — Contactless integrated circuit(s)
                           cards — Proximity cards
                           Part 1: Physical characteristics

ISO/IEC 15693-1/2   Identification cards — Contactless integrated circuit(s)
                           cards — Vicinity cards
   ISO/IEC 15693-1   Part 1:   Physical characteristics
   ISO/IEC 15693-2   Part 2:   Air interface and initialization

ISO/IEC 14443 is the standard for contactless smart cards with an operating frequency of 13.56 MHz. ISO/IEC 15693 is the standard for contactless smart cards with a reader working range width up to 1 m.

ISO/IEC 10536-1 defines the physical characteristics for contactless smart cards, which essentially correspond to the contact smart cards.

ISO/IEC 10536-2 determines the dimensions and position of the antenna. Here two types of contactless transmission are defined: inductive and capacitive transfer.

ISO/IEC 14443 determines the characteristics for contactless smart cards with a working band by 13.56 MHz ± 7 kHz. For these smart cards the definition for the position of the antenna according to ISO/IEC 10536 is not valid. Regarding the physical characteristics of the card, the standards of the contact smart cards are used in a general manner and adapted.

ISO/IEC 15693 determines the characteristics for contactless smart cards with a reader distance width up to 1 m.

1.2.3 Applications

The contactless smart card can be used favourably in those areas where communication from the reader with the smart card must take place in the shortest time, i.e. if where the inserting of the card into a reader is consuming too much time; for example, the case when entering public traffic systems. The contactless smart card system enables the possibility of a passenger walking through the gate and the system releases the gate automatically, provided the passenger has a valid ticket.

There are many other possible applications for contactless smart cards, e.g. conditional access to an office building or a football stadium or for purchasing in an automated department store.

Figure 1.13 Part-transparent contactless smart card (source: Giesecke & Devrient)

The contactless smart card shown in Figure 1.13 is displayed part-transparent in order that the functional electronic parts become partially visible. In practice, the chip module and antenna are located inside the card body and covered with opaque foils. This is also emphasized as a special advantage of the contactless smart card because in contradistinction to the contact smart card no part of the smart card surface needs to be reserved for the chip module and the entire card surface is available for the card design.

But, there are contactless smart cards which deviate from this rule.

An example is the contactless smart card with a chip module, which has an etched antenna placed on the module carrier where the chip inside the module is contacted with the antenna. This type of contactless smart card is called a transponder chip module or 'antenna on module'.

The structure of this card corresponds to the contact smart card which is represented in Figure 1.1.

For production of this smart card therefore the same manufacturing plants can be used as for the production of the contact smart card. This type of smart card can be used, for example, as a calling card.

With this system the transponder chip module smart card is presented on the surface of a reading device and connected contactless.

The advantage of this contactless system is that the telephone device does not need a contact smart card reader which is susceptible to vandalism.

1.3   Smart cards with multiple interfaces

The capability to be able to use a contactless smart card in contact smart card systems lead to the integration of both systems in a card, thus the following different systems are implemented:

1.3.1   Hybrid smart cards

Figure 1.14 shows the structure in principle of a hybrid card. This card contains a contactless chip, which is inserted in a chip module and attached to an antenna. Additionally to the contactless functionality a contact chip module is embedded in the smart card. The contactless chip and the contact chip function independently in different systems. Communication between the two chips can be undertaken by a reading device which contains both systems (hybrid terminals).

1.3.2   Dual-interface smart card

Figure 1.15 displays a dual-interface card schematically. This card contains a chip which is connected electrically to both systems through the outside contacts of the contactless chip module and through the antenna. A chip with combined contactless and contact function was developed particularly for this application.

1.3.3   Applications

The hybrid card is a multi-functional card and offers the possibility of integrating the function of the contactless smart card in an existing contact card system.

Typical applications are combinations of conventional smart cards (identification cards, bank cards and calling cards) with typical functions of the contactless smart cards (tickets for sports events and for public transport systems).

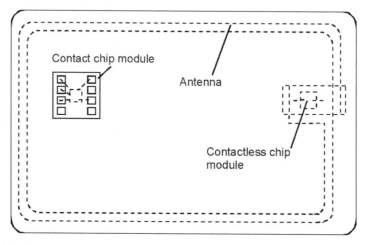

Figure 1.14 Hybrid smart card with a transponder with a contactless chip and an additional contact chip module

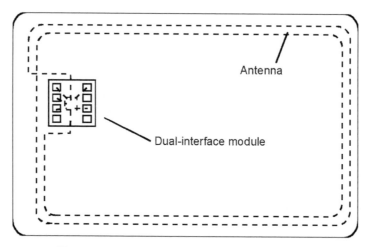

Figure 1.15 Schematic drawing of a dual-interface card

The dual-interface card is likewise a multi-functional card, whereby the integration of the contact and the contactless function on a chip offers the possibility of using multi-functionality within one system.

A typical application for the dual-interface smart card is the multi-functional cash card. With this smart card the owner can load an amount in his/her electronic purse in a secured environment like an ATM at the bank. Through the contacts the customer then has the possibility to make electronic money transactions with contact applications and/or, for example, pay for a train ticket via the contactless functionality.

2 Card body for smart cards

Many people today have a credit card, an identification card or other plastic cards in their pocket.

Indeed today most common cards are quite complicated technical devices, whose production requires the control of various techniques. This chapter is an overview of the substantial items and characteristics of standardized cards.

2.1 History

The identification card belongs to the category of portable falsification-safe documents which have a long history.

In its current form, however, the identification card found widespread use for the first time in the 1950s in the USA. There, one evening in February 1950, Frank MacNamara came up with the idea of the credit card after he could not pay his dinner in a restaurant because he had forgotten his wallet. In the same month, on 28 February 1950, he created the Diners Club together with some influential friends.

The first credit card issued by the Diners Club was made from cardboard. Very soon it became a status symbol and gained many members. In the year of foundation the club had 10,000 members and 1000 contracting partners. A worldwide expansion of the club followed.

In 1958 the international active travel and financing enterprise known as American Express decided to introduce its own credit card.

At the end of the 1950s the Bank of America created a credit card which was different from the predecessors by the fact that it also assigned licences for the issuing of this card to other banks and credit institutes. Following this model later worldwide VISA, MasterCard and EUROCARD were produced.

PVC (polyvinyl chloride) was used as card material due to its excellent optical and mechanical characteristics. The card became a full plastic card. The Eurocheque card, which was introduced for the first time in 1968 by European banks and credit institutes, was however a paper-laminated card. It consisted of a specially developed security paper with a security string in watermark paper which was in-laminated between PVC foils. The PVC material was also the base for standardization of the card. On the basis of experience with PVC cards, characteristics were determined which defined the card as a flexible, portable, falsification-safe document.

2.2 Materials

Card materials are supplied as foils for laminating or as granulates for injection moulding of the cards. Card materials are usually no-standard materials and have to be specially adapted for this application. They must fulfil many special criteria for the respective application of the cards. The following lists some typical criteria:

- Possibility for lamination and injection moulding
- Accuracy to size
- Printability
- Optical quality and colour constancy
- Mechanical firmness
- Lifetime
- Thermal stability
- Humidity resistance
- Solvent stability
- Anti-statics
- Physiologically harmless
- Environmental compatibility during production and use

The PVC material, which has been used for many years as card material, is today partly displaced by alternative materials due to new criteria, e.g. better environmental compatibility, longer lifetime and higher temperature resistance. The following is a quick overview of some different card materials as a rough comparison:

PVC (polyvinyl chloride)

Advantages:
- Low material price
- Experience of use over many years and adjustment to the processes of the preparation of cards
- Recycling possible

Disadvantages:
- Environmental compatibility
- PVC contains chlorine. Therefore PVC is hardly inflammable and by non-optimal combustion produces dioxins and furane, which are highly toxic
- PVC has limited thermal stability

Vicat softening temperature: depending upon type 65—80 °C

PC (polycarbonate)

Advantages:
- High temperature stability
- High mechanical strength
- Recycling possible

Disadvantages:
- Material is high in price
- Low resistance against scratching

- High heat consumption when laminating, difficult handling (during e.g. lamination and card punching)

Vicat softening temperature: approx. 150 °C

ABS (acryl butadiene styrene)

Advantages:
- Specially suitably for injection moulding
- Temperature stability
- Recycling possible

Disadvantages:
- ABS does not fulfil the ISO standard (for PVC) which describes non-inflammability
- In contrast to PVC, ABS is chlorine-free; however, ABS is not classified as environmentally friendly

Vicat softening temperature: approx. 100 °C

PET (polyethylene terephthalate)

PET is well known under the trade name polyester. Crystalline PET is very thermostable and has high mechanical stiffness — it is the best material for laminating. PETG, a glycol polyethylene terephthalate, modified for smart card production is a very good variation of PET.

PETG (polyethylene terephthalate glycol-modified)

Advantages:
- Most suitable material regarding environmental compatibility
- Economic (nevertheless not to the same extent as PVC)
- Recycling possible

Disadvantages:
- Presently there are not as many additives for the finishing of PETG incomparison to those available for PVC
- With some processing parameters PETG has a tighter process window than PVC

Vicat softening temperature: approx. 70 °C

Paper or cardboard

Advantages:
- Paper is an environmentally compatible material, as it is made of regenerating material (cellulose), in contrast to plastics which are made of crude oil — a raw material with restricted reserves
- Cards, made of paper, are already cheaper today than plastic cards. The price difference will increase further when the reserves of crude oil diminish
- Knowledge from the paper processing industry could be used
- Cardboard cards could be very thermostable
- Recycling is possible and not unusual

Disadvantages:

- Cardboard cards are very sensitive to humidity, even air humidity can influence the accuracy of the size of the cardboard card negative
- Cardboard cards require careful handling. They are sensitive to abrasion, folding, edge impact and other mechanical loads. They do not fulfil all (formulated for PVC cards) ISO standards

Taking account of the disadvantages mentioned above it is expected that cardboard cards will only become established for certain applications. Typical applications include the case of short-life throwaway cards and contactless smart cards, which do not have to be put into the slot of a contact reader.

Additional materials

In additional to the card materials mentioned there are other materials available on the market or in development. For example, there are telephone cards available which are made of wood; in addition, other plastic materials such as polypropylene or biologically degradable plastics. These new materials have yet to prove themselves in field trials.

2.3 Standards

The most important standards for the smart card have already been listed in Section 1.1.2. Standards particularly valid for the card body are:

ISO/IEC 7810	Identification cards — Physical characteristics
ISO/IEC 7811-1/2/3/4/5/6	Identification cards — Recording technique
ISO/IEC 7811-1	Part 1: Embossing
ISO/IEC 7811-2	Part 2: Magnetic stripe
ISO/IEC 7811-3	Part 3: Location of embossed characters on ID1 cards
ISO/IEC 7811-4	Part 4: Location of read-only magnetic tracks — Tracks 1 and 2
ISO/IEC 7811-5	Part 5: Location of read-write magnetic track — Track 3
ISO/IEC 7811-6	Part 6: Magnetic stripe — High coercivity
ISO/IEC 7812-1/2	Identification cards — Identification of issuers
ISO/IEC 7812-1	Part 1: Numbering system
ISO/IEC 7812-2	Part 2: Application and registration procedures
ISO/IEC 7813	Identification cards — Financial transactions cards

2.4 Card elements

The different components of a card are termed card elements. Card elements which help to ensure the security of a card are called authenticity elements. Authenticity features of the card can partly be detected by people without the help of machines. These authenticity elements are called humane authenticity elements in contrast to machine-readable authenticity features, whose recognition is only possible by special devices or machines.

In the following the most important card elements are presented:

2.4.1 Plastic foils

As previously mentioned, the production of cards from plastic foils is the most common manufacturing method. Usually several foil (sheets) are stacked on each other and laminated. In the laminating process the stocked sheets are clamped between two heatable plates. The sheets are heated until the softening point of the plastic is reached. Afterwards, the foils cool down in the clamped position (also see section 3.1). This process connects all single foils together to an homogeneous sheet which corresponds to the card thickness. In a further manufacturing step, the cards are punched out.

The foil material can be, for example, PVC, ABS, PETG or PC. It is also possible to mix different foil materials or foils from mixed materials (so-called blends, for example, 60 % ABS / 40 % PVC blend).

At the present time cards from PVC foils are very common. It is expected that for reasons of environmental protection or for a higher temperature resistance or a higher lifespan of the card alternative materials besides PVC will be used in the future, e.g. PETG, ABS and PC.

The following describes the most important characteristics of the foils used for card manufacturing.

Foil thickness

The thickness of the used foils depends predominantly on the number of plastic foils and the level of thickness gradations used for the production of the card body. The thickness tolerances of the used foils have to be coordinated in a way which allows the total thickness of the manufactured card body to stay in the required ISO standard. As already described in Section 1.1, the actual demand for the card bodies used for the production of smart cards is:

$$k = 800 \pm 40 \, \mu m$$

Where k is the required card thickness.

This demand is based on the available thickness of chip modules (max. 580 μm) on the one hand and on the other hand on the tendency to arrange the remaining wall thickness in the card cavity (cavity in the card for embedding of the chip module) to be as thick as possible in order to withstand mechanical overloads in daily usage.

The thickness of the foils, which are used for the production of the card, can be calculated from the following formula:

$$f_1 + f_2 + ... + f_N = k + s$$

Where $f_1, f_2..., f_N$ are the thicknesses of the individual foils and N is the number of foils used. This formula expresses the fact that the thickness of the laminated card is thinner than the total thickness of the single foils. This difference in thickness, s, is the so-called shrinkage of the plastic material. The shrinkage is not a material loss, it is only a thickness loss.

The loss s depends on the foil material and on the number of foils, as well as the roughness depth and roughness structure of the foil surface and the measuring instrument and the measuring method for determination of the foil thickness. The loss s also modestly depends on the laminating technique and the laminating parameters. The loss s is experimentally determined in each case, thus that the foil thicknesses f_1, f_2 ..., f_N are measured before laminating at the same measuring point (for example card centre), and after laminating the card thickness k at the same measuring point is likewise determined and from that the shrinkage s is calculated.

Thickness tolerance of foils

The required thickness tolerances of the finished card as already mentioned above, are mostly +/- 40 µm. That corresponds to +/- 5 % of the card thickness. In order to maintain these total tolerances in all produced cards, it is necessary that the supplied foils are in a range of tolerance within a maximum +/- 5 % of the foil thickness. With thinner foils this is however not always possible. The thickness tolerances for thinner foils (for example 150 µm) are +/- 7 % or with very thin foils (for example 60 µm) +/- 10 % of the foil thickness. Therefore with cards which are structured functionally of several thin foil layers a statistical thickness calculation is executed. This calculation is based on the fact that the probability that at one point of the card all foils indicate the maximum or the minimum thickness at the same time is infinitesimally small and that this probability decreases with the increase of the number of used foils.

If one presupposes a normal distribution for the statistical distribution of the thickness tolerances of the foils, then one can use the following approximation formula for the statistical sum tolerance (also see Section 4.2.1, statistical calculation of the rear wall thickness):

$$\Delta k^2 = \Delta f_1^2 + \Delta f_2^2 + \Delta f_3^2 + ... + \Delta f_N^2 + \Delta s^2$$

Foil colours

In principle there are transparent and opaque foils. Opaque foils are usually white and mostly used as inlay. The white tone of the foil must be agreed upon with the foil manufacturer and held by reference samples. The foil colour can be defined additionally by colorimetry according to DIN 5033. With the transparent foils, which are mostly used as overlay foils, the light permeability can be agreed upon according to DIN 5036 for visible light or ultraviolet and/or infrared light.

Further it is of interest to define the stability of the white foils against bleaching by influence of light. That is defined in the standard DIN 54004. It is also important to check the complete smart card and not only the single foils against sun radiation influence regarding bleaching. Bleaching influences the entire optical impression of the smart card and can cause both aesthetic and security-relevant problems. It is also a safety-relevant issue because a common continuous optical impression offers the possibility of detecting deviations in the appearance of the card as falsification.

Thermostability

A measure for the thermostability of a foil can be defined by the Vicat softening temperature according to DIN/ISO 306. Section 2.2 shows typical values of the Vicat softening temperatures for the most important card materials.

The Vicat softening temperature is additionally an orientation value for the necessary laminating temperature.

Foil surface

An important characteristic of a foil is the roughness depth of the surface and the texture of the surface. The roughness depth of the foil surface is defined according to DIN 4768.

Further the printability of the foil surface is an important characteristic. The printing technique must be agreed upon with the foil manufacturer and the printability must be guaranteed.

Further characteristics

Further foil characteristics of interest for card production are:
- Tensile strength according to DIN 53455 (important for mechanical stability)
- Shrinkage according to DIN 53377 (important for accuracy of size after the laminating process)

2.4.2   Printing

In principle it should be differentiated as to whether the printing is executed on the outside surface of the card (external printing) or on an inlay of the card (internal printing), which is covered with a transparent overlay foil and laminated.

The internal printing is the most common type of card printing and its advantage is that the printing surface is well protected from abrasion. If design printing is done under an overlay foil a problem could occur: The printing behaves as an interface between the overlay foil and the foil with the printing inlay. Therefore adhesion of the two foils is reduced. The reduction of the adhesion is reduced when the thickness and density of the printing increases. There are two possible solutions to avoid the reduction of adhesion. The first is to use modified silk-screen inks, which are optimized for the lamination process. The second solution is to use adhesive-coated overlay foils. In this case the adhesive makes the connection to the artwork printing. However, with offset printing surfaces it must still be noted that areas with high colour density cannot be optimally laminated by use of adhesive-coated overlay foils. The limits of the colour density have to be found using experimental tests.

Credit cards and bank cards are usually manufactured with printing under a transparent overlay. Different printing techniques are used for their production, e.g. screen printing, indirect letter printing, wet offset and drying offset. Also different techniques of safety printing are used, e.g. Gouillochen printing. These are complex curved fine lines, as they are mostly implemented in banknote printing.

External printing in high volume was used for the first time with the German calling card. In order that the chip module built in the card was backlaterally covered, a reversal of the so-far-usual card construction was made. The core of the card consisted of transparent foils, and these external foils were opaque and were printed. For the purposes of cost reduction this structure was modified later.

For protection of external printing a transparent protective varnish is printed upon the artwork. Thus, protection against abrasion and other wear is not as high as with internal printing which is covered with an overlay foil. External printing permits huge possibilities of transferring a fastidious printing-technique in an economical way as the restrictions with high colour density printing are not as marked as with internal printing.

External printing in ultraviolet (UV) offset is mostly covered with a UV protective varnish.

The most important compression matters for the printing of cards are represented in the following:

Offset

A process termed photolithography is used to make offset printing plates. In the first step a positive film or a negative film depending upon the offset system used is made from the image. As a next step the negatives are exposed on a thin metal plate (aluminium or zinc) which has a light-sensitive coating. Light from powerful lamps shines through the negative and hardens the image on the plate. The plate is developed and then chemically treated, some areas of the image on the plate take the ink and some areas repel it. This effect occurs when the plate is made wet by a dampening roller. Depending on the offset system the water used could be distilled water of a special mixture.

The turning plate cylinder then offsets the inked images onto the rubber blanket cylinder. The rotating blanket cylinder, in turn, offsets the image onto the foil carried by the pressure roller. Figure 2.1 shows the principle of wet offset printing.

Additional to wet offset printing there is dry offset. The printing plates for dry offset are prepared so that only the image will accept ink during the contact to the ink rollers.

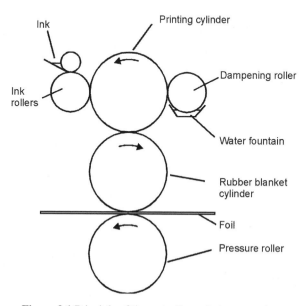

Figure 2.1 Principle of the wet offset printing procedure

Printing inks

If part of a full-colour image is enlarged it can be seen that the image is only an accumulation of many small dots of different colours. These different colours are the three primary colours and additional black. The three primary colours are:

CYAN
MAGENTA
YELLOW

These colours are termed primary colours because they cannot be made by the mixture of other colours, while all other colours can be mixed from these primary colours. In the following some examples of colour mixtures are mentioned:

CYAN   + MAGENTA = VIOLET
YELLOW   + MAGENTA = RED
YELLOW   + CYAN   = GREEN

The mixture of the three primary colours results in black, which is, however, not a perfect black colour and appears as dirty black. For this reason in addition to the three primary colours with high-quality prints the black colour is also printed.

In previous times for production of negative or positive film for offset printing the image was striped down to every individual primary colour and black so that there were four printing plates. Nowadays, however, the separation of the image into the primary colours and black is done with a colour scanner and a computer. The collected colour information can be transferred directly to the printing machine. The information in the printing machine is transferred with a laser beam directly to the printing plate.

With this method it is also possible to modify the printing quality without stopping the printing machine.

Screen printing

The equivalent of the printing plate from offset printing is, for the screen printer technique, the screen — a wooden or aluminium frame with a fine nylon mesh stretched over it. The mesh is coated with a light-sensitive emulsion or film which — when dry — will block the holes in the mesh. The image that needs to be printed is output to film either by camera or image-setter. This positive film and the mesh on the screen are sandwiched together and exposed to ultraviolet light in a device called a print-down frame. The screen is then washed with a jet of water which washes away the light-sensitive emulsion that has not been hardened by the ultraviolet light. This leaves you with an open stencil which corresponds exactly to the image that was supplied on the film.

The screen is fitted on the press and is hinged so it can be raised and lowered. The foil to be printed is placed in position under the screen and ink is placed on the topside of the screen. The frame also acts as a wall to contain the ink. A rubber blade gripped in a wooden or metal handle called a squeegee (not unlike a giant windscreen wiper) is pulled across the top of the screen; it pushes the ink through the mesh onto the surface of the foil. A pressure roller presses against the screen to bring the foil in close contact with the mesh. To repeat the process the squeegee floods the screen again with a return stroke before printing the next impression. Figure 2.2 shows the principle of screen printing.

The printed foil must be led through a drying tunnel afterwards to dry the printing, before it can be stacked.

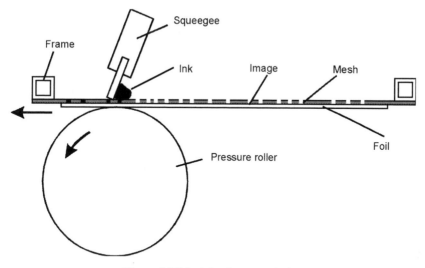

Figure 2.2 Principle of screen printing

Gouillochen printing

Gouillochen are the fine, harmonious lines of patterns present on bank notes. They are recorded through specially designed Gouillochen machines. The Gouillochen machine has a recording device which can record the oscillations of a swinging body on a glass plate. It has a multiplicity of mechanisms by which different overlaid oscillations can be produced, one separated curved line structure is recorded in each case. By direct absorption of the oscillation the Gouillochen machine draws neighbouring curve families which result in the total Gouillochen.

It is also possible to simulate these oscillations using a computer program and allow the computer to produce the Gouillochen.

A Gouillochen manufactured in such a way is then built into the layout and transferred onto the printing plate. Usual printing techniques for Gouillochen printing are:

- Low steel pressure
- Indirect letter press printing
- Offset printing

Offset printing has already been described.

Indirect letter press printing

Letter press printing or typographic printing is a way of printing where the ink is transferred to raised structures of the printing plate and then transferred to the plastic foil. The printing plate can consist, depending upon the machine type, of corroded plates or also of chemical lithographically produced plates with raised structures.

With indirect printing the colour will not transfer directly from the printing plate to the foil, but first to a rubber blanket cylinder and then from there to the foil.

2.4.3   Signature panel

It is possible to place a signature directly on a card surface with special foil pins. However, for practical reasons the demand is for the signature on the card to be made possible with commercial ballpoint pens. A prerequisite for this is that the signature field has a gripping surface which binds the ink of the ballpoint pen in such a way that the signature cannot be wiped away. There are different forms of signature panels which are described in the following:

Paper signature panel

A possible solution is the paper signature strip: Before laminating a paper panel is mounted on the outside overlay foil of the card. It connects itself to the card surface during the lamination process. The signature panel can be a security paper with watermarks and can contain additional security features which protect the signature against counterfeiting; for example, signature panels which react to solvents. If somebody wants to alter the signature the colour of the signature panel changes.

Advantages:
- High safety level against counterfeiting
- Good possibility to write on

Disadvantages:
- Reduced lifespan by wear out and abrasion of the paper
- Deformation of the card by different thermal expansion of paper / plastic during the laminating process

Screen printing signature panel

With special colours which contains rough and absorbent pigments the signature field is mostly printed on the outside overlay foil in white silk-screen printing. Following this, on the white surface additional security printing is printed using offset printing technology. The so-prepared overlay foil is accurately mounted onto the remaining foils and then laminated. Also with the screen printing signature field it is possible to integrate security features in the screen printing.

Advantages:
- High lifespan
- Economical production

Disadvantages:
- Overlay foils are often very thin and hard to handle in the printing process and when assembling the sheets stick together before laminating.

Hot-stamp signature panel

The hot-stamping procedure is often used during the preparation of smart cards. The principle of the hot-stamping procedure consists of transferring a prepared surface from a carrier tape to the card surface by usage of a heated stamp.

Figure 2.3 shows the principle of hot stamping. With a hot stamp a stamped tape is embossed on the surface of a prefabricated card. The stamping tape consists of a carrier tape which is heatproof in the range of the embossing temperature. The back side of the carrier surface is prepared in such a way that at room temperature a sufficient adhesion for various printing cycles is maintained at the seal temperature no matter how the adhesion is reduced. This surface is printed with a mirror image layout and printed in reverse order of the printing cycles, so that the signature stamp is arranged under the carrier form. The signature panel surface is covered with a heat-activatable adhesive. When putting the heating stamp on top of the stamp tape and card surface the signature panel connects itself via the heat-activatable adhesive with the card surface and removes itself from the carrier tape.

Advantages:
- Rejects during smart card production are reduced because of removal of the signature panel from card production
- The laminated card without a signature field has a lower safety value. An automatic counting of the single card in the signature embossing station increases the total security in card production

Disadvantage:
- High manufacturing costs of the stamp tape. The process is not suitable for small editions of the stamping tape

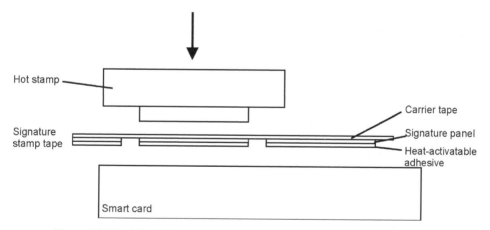

Figure 2.3 Principle of the hot-stamp method for a signature panel on a smart card

2.4.4   Magnetic stripes

In the following international standards the technical characteristics of the magnetic stripe are determined:

ISO/IEC 7811-1/2/3/4/5/6   Identification cards — Recording technique
    ISO/IEC 7811-2   Part 2:   Magnetic stripe
    ISO/IEC 7811-4   Part 4:   Location of read-only magnetic tracks —
                             Tracks 1 and 2
    ISO/IEC 7811-5   Part 5:   Location of read — write magnetic track —
                             Track 3
    ISO/IEC 7811-6   Part 6:   Magnetic stripe — High coercivity

Testing methods for cards with magnetic stripes are determined in standard ISO/IEC 10372-2.

In the following some examples are picked out from the standards. For further details the reader should refer to the appropriate standards.

- The magnetic stripe which is defined in these standards carries no additional protective layer on the magnetic material
- The area for magnetic material for tracks 1, 2 and 3 is defined as in Figure 2.4
- The magnetic stripe has three tracks, in which data are coded in the form of locally magnetized bar-shaped areas. The density of the codable information for each track is as follow:

Track 1   8.27 bits per mm (210 bits per inch)
Track 2   2.95 bits per mm (75 bits per inch)
Track 3   8.27 bits per mm (210 bits per inch)

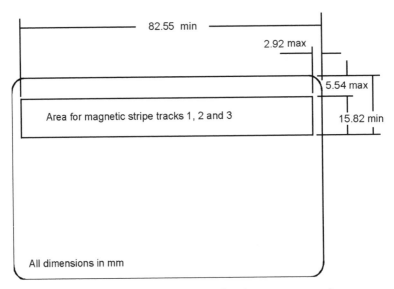

Figure 2.4 Position of the magnetic stripe on a smart card

- The position of the three single tracks is shown in Figure 2.5
- ISO/IEC also defines smart cards with only two magnetic tracks, 1 and 2. The area for magnetic material can be smaller in this case and is defined
- The characteristic of the magnetic stripe is compared with a reference magnetic stripe which is defined in ISO/IEC 7811-2. Reference cards with reference magnetic stripes for calibrating of the measuring instruments are available at the Federal Standards Laboratory
- ISO/IEC 7811-6 defines a high coercitive magnetic stripe. The coercitive describes the strength and resistance of the magnetic fields of the coded areas on the stripe, and is defined as magnetic field strength which is necessary in order for the saturation magnetized magnetic stripe to demagnetize completely (to the field strength 0). The unit for coercitivity is the unit of the magnetic field strength oersted [Oe]. For magnetic stripes the standard does not specify the number of the coercitivity. It only wants comparative measurements with the determined reference magnetic stripes. In practice, the magnetic stripe according to ISO/IEC 7811-2 has a 300 Oe magnetic stripe (so-called lowco) and the magnetic stripe called highco according to ISO/IEC 7811-6 has a 4000 Oe magnetic stripe.

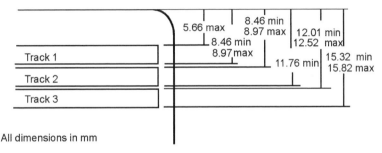

All dimensions in mm

Figure 2.5 Position of the three data tracks measured from the top-right edge of the smart card

Generally there are two possibilities for getting a magnetic stripe on the surface of a smart card.

One is that the magnetic stripe is applied on every single card body and the second possibility is that the magnetic stripe is applied over the complete plastic sheet.

With single tracking the card initially is manufactured without a magnetic stripe. Afterwards, the magnetic stripe is transferred to the card surface using the hot-stamping technique. A prerequisite is that the card surface is suitable for this procedure. Figure 2.6 shows the procedure principle for single tracking.

A magnetic stamping tape, consisting of a thermostable carrier tape coated with a ferromagnetic material and additionally with a heat-sealing layer, is pressed on the card surface with a hot-stamping roll. The magnetic layer connects itself with the heat-sealing layer on the card surface and the carrier tape separates from the magnetic layer when it is taken off.

The second way to make a smart card with a magnetic stripe is to apply the magnetic stripe on an overlay sheet as shown in Figure 2.7. A magnetic stripe, as wide as the plastic sheet, is obtained with a heated stamp on the surface of the overlay foil.

The overlay foil with the mounted magnetic stripe is stack over the other plastic foils. This stack is laminated and in the next production step the single cards are punched out of the sheets, which now have individual magnetic stripes.

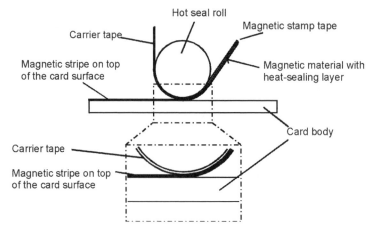

Figure 2.6 Single stamping of magnetic stripes on single card bodies

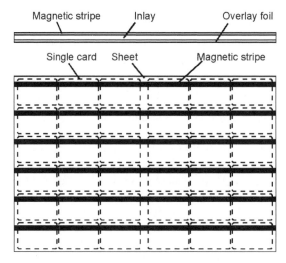

Figure 2.7 Applied magnetic stripes on a plastic sheet

2.4.5   Embossing

Embossing is a well known form of inscription on credit cards, where the name, credit card number and the expiry date is embossed. This embossed information can easily be transferred to the receipt through a ribbon.

Security printing is used to accommodate the embossed characters on a card surface. Counterfeiting attempts by modification of the letters or numbers leads to visible modifications of the overall view and can be detected by visual observation.

An advantage of embossing is that the dealer can handle the card system without large expenditure of devices and infrastructure. The disadvantages are: First, the embossed card becomes thicker. In an age where many people carry several cards in their wallet, the higher thickness of the card is a disadvantage for the card owner.

Second, the usage of stand-alone devices which only use the embossed information without online verification with the bank, if the credit card is still valid, is becoming a growing security problem.

Electronic devices are introduced to the field in a step-by-step manner. These devices offer the possibility of reading the card information from the magnetic stripe and/or from an embedded chip in the smart card. The online verification of the information is also very easy. In addition, the receipt is automatically outprinted. Some systems already do not use embossing any longer because of security reasons and have switched to the electronically stored information on the magnetic stripe or chip. However, in any case, in the immediate future embossing, a magnetic stripe and a chip will exist side by side on the smart card because the operators and issuers of the card systems require worldwide distribution of their card system, hence the card must function equally in differently developed infrastructures.

Embossing of the card takes place via deformation at low temperatures of the card material as shown schematically in Figure 2.8.

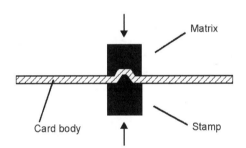

Figure 2.8 Principle of embossing of a smart card

The embossing tool consists of two parts: A matrix, which contains the character to be embossed as recess, faces a stamp, which contains raised characters. With open tools the card is led between the two tool parts. Thereafter the tool closes at room temperature whereby the stamp in-presses the card material into the matrix. The matrix which is necessary for the whole characters and the appropriate stamps is in each case in the perimeter of an embossing wheel, which is turned until the correct pair of tools for embossing are in the right position.

Standard ISO/IEC 7811-3 determines the position of embossed characters on the card. Figure 2.9 shows these definitions schematically.

The upper field with only one line is intended for the embossing of the identification number and the lower field with four lines is for the embossing of the name, the address and other information. The dimensioning of the characters refers in each case to their centre lines. The dimensioning of the fields refers to the outer contours of the characters. The characters represented in the drawing only serve for the explanation and do not correspond to valid definitions.

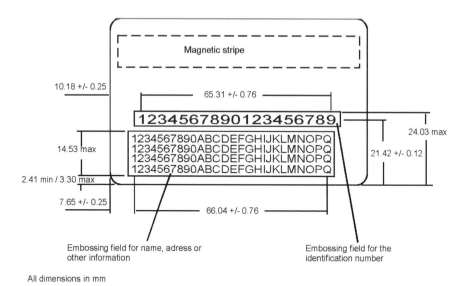

Figure 2.9 Position of the embossed characters on a smart card

Standard ISO/IEC 7811-1 determines the characters which can be used for embossing the card and their formatting, and their relative heights as well as demands on the imprint of the embossed card. The relative height of the embossed characters is determined with visual and machine-readable characters as follows:

Relative height: 0.48 +/- 0.08 mm

The embossing of the card leads to a deformation of its external dimensions. Standard ISO/IEC 7810 extends the dimensional tolerances of the embossed card as follows:

Mass of the embossed card (mass of the non-embossed card):

Width: 85.47 mm min. / 85.90 mm max. (85.47 mm min. / 85.72 mm max)
Height: usually 53.92 mm min. / 54.18 mm max (53.92 mm min. / 54.03 mm max)

Normally the embossed characters on the card are dyed in a manner suitable to the card layout, consequently they are well readable and the total optical appearance of the card is improved. A hot stamp prints a thermal transfer tape (similar to the hot embossing tape described above) on top of the embossed characters. Thus the ink connects itself with the surface of the embossed characters and separates from the thermal transfer carrier tape.

2.4.6   Laser engraving

The function of laser engraving is the local burn of foil material by a laser beam. The beam of a solid laser (Nd:YAG) is moved on the card surface, whereby each laser pulse burns a black dot into the card. With the appropriate grid of dots different grey tones can be made, which result in a high-contrast picture. Figure 2.10 shows this process schematically.

The laser beam penetrates through the transparent overlay foil and the energy is absorbed on the surface of the opaque inlays. In the figure, on the right side of the laser beam a laser-engraved dot is shown. As evident, the burn of the foil material begins at the surface of the inlays, goes through the transparent overlay foil away up to the surface of the card and is tactile like a welding spot. Both the overlay foil and the printed surface of the inlays are irreversibly changed through the laser engraving. This is a big advantage of laser engraving in relation to other inscription and printing methods for smart cards. Laser engraving has a typical appearance, is tactile and penetrates deeply into the card and into the inside laying print which is an advantage for the security of a smart card.

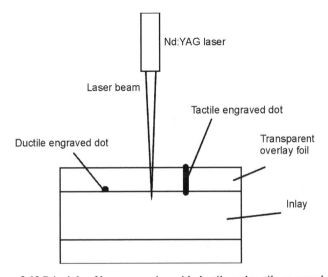

Figure 2.10 Principle of laser engraving with ductile and tactile engraved dots

For optimal function of laser engraving an accurate tuning of the foil materials in respect of the procedure is necessary. The transparent overlay foil is doped with special additives which favour the burn and the change of colour by the burn. The laser limits the burn locally and the resultant gases blister the foil at the surface of the card. Here it is made certain that the dots are closed.

For some applications laser engraving should not be perceptible on the card surface. With the right focusing of the laser beam and tuning of the card construction it is possible to prevent the laser engraving penetrating up to the surface of the card.

Using laser engraving a photo can be engraved on the card and at the same time the card can be personalized. For engraving a photo the laser beam must step through the entire image and produce the picture in the form of a raster.

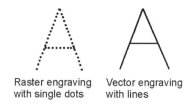

Raster engraving   Vector engraving
with single dots   with lines

Figure 2.11 Difference between raster and vector laser engraving

The raster lasers can also be used for writing the characters on the card, see Figure 2.11. It is conceivable that the laser beam has to scan through a rather large area which accordingly needs a long processing time. For this reason an alternative to the raster engraving, a rational procedure for personalization was developed, the so-called 'vector engraving'. With this procedure the laser beam goes after the outline of each character like a pencil during writing and can engrave alphanumeric characters in a much shorter time.

2.4.7   Hologram

The function of the hologram is based on the wavy expansion of light. From physics it is well known that under certain conditions standing waves can develop through overlay of two or more waves. This effect can be shown experimentally with waves in water. Using a flat container in two places a punctual vibrating of the water surface is caused by pulsating compressed air. This produces centric waves of the same frequency and amplitude. Figure 2.12 shows these waves as wave 1 and wave 2 by concentric circles. At the intersection of both circles an overlay of the two waves develops which runs parabolic. Standing waves are formed along these parabola.

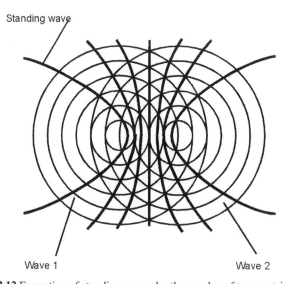

Figure 2.12 Formation of standing waves by the overlay of two centric waves

The formation of a three-dimensional (3D) picture by a hologram can be compared with this experiment, where the hologram produces a similar effect – only in three-dimensions. Many light waves overlay in such a way that a 3D wavefront develops which produces the 3D picture.

3D holograms by their fine structures produce a light wavefront when being lit with a laser. The hitting light is reflected in such a way by the structures of the hologram that it is assumed as a second source of light with a phase shift in relation to the hitting light. From the overlay of the two light waves results the light wavefront which looks as if it was reflected by a 3D object.

In addition, it is possible to produce flipping pictures by the technique of holography. Here, an optical feature is received, which supplies different picture information depending upon the viewing angle.

Most holograms which are used on smart cards are white-light holograms. This means that they can fulfil their optical function with regular daylight and a laser or another special source of light is not necessary to see the object in the hologram.

The structures of a hologram are wavy in the micrometre area, and indicate no similarity with the produced picture. They work to bend the incident light in many different directions so that the desired light wavefront develops.

The hologram structures are manufactured in production by hot stamping and metallizing of foils. The manufactured hologram foil is coated on the back side with hot adhesive and covered on the front side with a protective plastic film. With another hot stamping, this time at lower temperatures than in the production of the hologram, the hologram is transferred off the protective plastic film onto the card's surface.

The shape, size and position of the hologram are not determined to general standards and are defined in customized specifications.

2.4.8   Multiple laser image (MLI)

The MLI is similar in one way to the hologram as it produces flipping pictures, but it differs by the fact that these tilting pictures are produced by laser engraving underneath the card surface. The card surface is provided with lens structures within the area of the MLI. These lens structures consist, for example, of staff lenses which are arranged next to each other. They are produced when laminating the card. Therefore the appropriated laminating plates carry profiles which form the lens structure when laminating the card and imprint them into the surface of the card.

Figure 2.13 shows the principle of an MLI.

The MLI is produced by laser engraving. The two tilting pictures are engraved closely next to each other with different angles of incidence for the laser beams. In the figure, dot **1** and dot **2** of two flipping pictures are shown while they are engraved by the laser beams **1** and **2**. The lens structure means that only figure **1** or figure **2** can be seen, depending on the viewing angle.

In contrast to the hologram, the MLI is an integral part of the card and therefore offers additional security against falsification. Since the tilting pictures of an MLI are produced directly by laser engraving inside the card, the MLI offers the possibility to engrave every individual card with a non-standard flipping picture, e.g. the signature of the owner or expire date.

This is also a method to increase security against falsification.

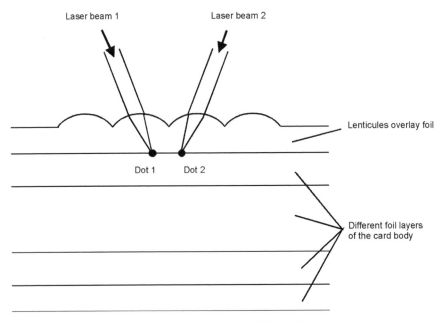

Figure 2.13 Schematic principle of an MLI (multiple laser image)

2.4.9   Additional card elements

In addition to the previously mentioned card elements there is an abundance of further card elements. Some of them are described briefly in the following.

Machine-readable features

Machine-readable features are additional authenticity features for the authenticity check in reading machines. A special feature material and a recognition sensor are part of the machine-readable features. The feature material is laid on the card materials, or added into the card body, and detected by readers with special sensors which are integrated in the reader. The high security of machine-readable features makes use of the fact that the feature material is a special development which is not available on the market for other applications.

The recognition sensor has a safety structure which does not permit the recognition of the physical measurement principle. If, for example, an ATM does not recognize these features it does not accept the credit card.

UV colours

For UV colour printing inks are used which modify their colour when illuminating with UV light or become visible and then light up in a typical colour.

Micro-printing

Texts and characters are printed with special printing techniques which are not recognizable by the naked eye and can only be seen with a magnifying glass.

Security strings

Security strings, which are also used in bank notes, can be implemented into smart cards. The strings are mounted on an inlay foil and covered with transparent overlay foils and can be seen afterwards from outside the smart card.

Bar code

The bar code is brought onto the surface of a smart card by e.g. ink-jet printing, thermal transfer printing or laser engraving and can be read by a machine. There are also two-dimensional security bar codes.

Optical memory

Here the technique of CD-ROM (compact disc read-only memory) is transferred to smart cards. There is one difference, the data tracks of optical memory cards are usually arranged parallel to the long side of the card and special readers are used for reading the data.

The optical memory card contains a reflective optical recording medium encapsulated between transparent (mostly polycarbonate), protective layers. The information is stored digitally in a binary code of '1' or '0' bits which are represented by either the presence or the absence of physical spots on the recording media. Using an optical memory card information up to 4MB can be stored.

Thermal transfer printing

Thermal printing technology uses a print ribbon that contains sublimation dyes suspended in a wax carrier. The low temperature of the wax thermal process transfers the sublimation dyes to the smart card surface, but does not sublimate them.

A thermal transfer printer works like a needle printer, whereby the printing needle is heated. The hot printing needle transfers the pixels from the ribbon onto the card surface. The technique allows printing of logos, photos and characters on the card surface with high quality.

Thermal sublimation printing

Sublimation is a transfer process which uses heat-sensitive sublimation inks to permanently dye substrates with polymer-coated surfaces. Sublimation describes the process of a solid substance changing directly into a gas or vapour without passing through the intermediary liquid state first. For printing on a smart card surface the term sublimation is used to describe heat-activated inks that change into a gas when heated and have the ability to bond with polyester or acrylic surfaces. Sublimated images are exceptionally scratch resistant because the image is protected within the surface.

A sublimation transfer is placed on the surface of a smart card. Using heat and pressure from a standard heat press, the inks are activated and begin to change into a gas. At the same time, the molecular chain of the polymer surface expands and forms openings which receive the individual dye molecules. Very high pressure ensures that the vapour is forced inside the card surface. At the exact moment that the substrate is removed from the heat the material cools and the polymer surface regains its original form. Sublimation gases within the surface will revert back to a solid state and remain trapped indefinitely within the surface of the smart card.

Optical variable ink (OVI)

OVIs are printing inks with which images can be produced that appear different colours at different viewing angles. For this technique the production of micro-caps is used. This technique enables the shrouding of colour pigments with very thin part-transparent layers in different colours.

OVI is a registered trademark of the printing ink manufacturer SICPA.

Figure 2.14 schematically shows the principle of OVI.

From the perpendicular view (direction **A**) mixed colours **1** and **2** of the two colours appear. Twisting the printed surface (direction **B**) changes the mixing proportion of the two colours for the viewer. The colour change, whereby then colour **1,** dominates.

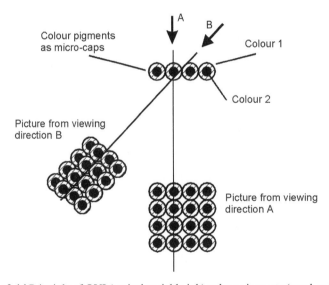

Figure 2.14 Principle of OVI (optical variable ink) colour pigments (much enlarged)

Nano-code

For this very thin dye films (layer thickness measured in nanometre = 1 millionth of a millimetre) in different colours and in particular order are printed on, one above the other. The one-above-the-other printed dye layers result in a film, which is then cut and grinded to a powder. Each pigment of this powder consists of the particles of the used dye layers. The authenticity of the printing can be checked at any time by analysis of the colour pigments.

Each pigment carries a particular coding by number of colours and order of the colour layering.

Gene code

There are already plans to use the hereditary molecule DNA (deoxyribonucleic acid) as an authenticity feature. DNA molecules are resolved in solutions and imprinted as characters on the card surface. The imprinted characters light up in particular colours under laser light.

3 Manufacturing processes for card bodies

As already described in Section 2.1, credit cards have been manufactured since 1950. Over the years new techniques were developed to protect the smart card against counterfeiting. The introduction of the magnetic stripe and the storing of data on it marked the entry of the machine readability of the card. For credit cards and bank cards visual security (e.g. holograms) is still very important. Since the smart card with its integrated circle (chip) was introduced into the market, for example, the telephone card and the GSM cards for mobile phones, there have been no more additional visual security features added to it. With these cards the card body, from a safety-relevant view, only fulfils the function of a carrier for the chip or chip module, respectively. Nevertheless, also with these simple cards a significant artistic expenditure is undertaken. One has to pay attention to fastidious design and high-quality printing, because the printing area can function as an advertising surface or is used for self-promotion of the issuing company.

While bank cards are mostly made of foils with heat lamination, as a lot of security characteristics require this technique, telecommunications cards are partly produced using an injection-moulding procedure. An overview of the different manufacture techniques for smart card bodies follows.

3.1 Laminating technology

Lamination is the connection of at least two plastic foils through simultaneously heating and pressing together. For the preparation of card bodies the laminating technique involves structuring a card of several plastic layers which are, for example, pressed between two laminating plates in a laminating press.

Before the single plastic foils can be brought to the laminating process the different foils have to be placed exactly above each other. If the adjustment is not done in a precise way it will cause problems when later out-punching the card body out of the laminated sheet. The problem occurs, if image **1** on the top overlay foil is not exactly above image **2** of the bottom overlay foil; the image on the back side will be cut off when punching out such a card (see Figure 3.1)

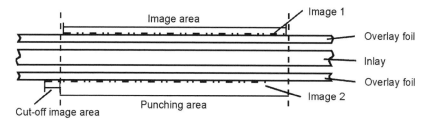

Figure 3.1 Misalignment of the image on the front and back side of a three-layer card

For exact adjustment of the two overlay foils different methods are possible.

The easiest way is to press the ledge of the three (or more) foils against stop pins. These so-adjusted single foils are fixed to each other together with a heated spot stamp in the periphery of the foil (see Figure 3.2). This adjustment method could also be achieved by a fully automated machine.

Figure 3.2 Sheet collating machine for the adjustment of single foils (source: Mühlbauer SSC 2500)

Another way is, if on the side of the foils some adjustment holes are punched out of a very thin single layer (see Figure 3.3). Afterwards the foils are stocked on the adjustment pins which are placed on a mounting table. This package of the single foil is also fixed with a heated spot stamp.

A more precise way of adjusting the foils against each other uses adjustment with printed crosses on both overlay foils. The adjustment crosses are printed at the same production step when the image printing is made (see Figure 3.3) and gives exact reference points.

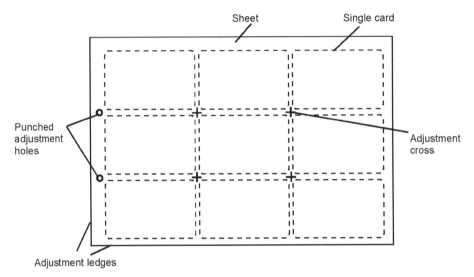

Figure 3.3 Different adjustment possibilities for printed sheets

With a special mounting table two camera systems are used — one camera is under the table plate and the other is placed above the table. The crosses from the bottom overlay foil and the top overlay foil are brought exactly above each other. Afterwards the adjusted foils are also fixed against each other with heated spot stamps.

Now the pre-mounted foils can be laminated.

While the layers are pressed together in the laminating machine they are heated up to the softening temperature of the plastic layers where the connection between the plastic foils is established. Afterwards the pressed layers are cooled down under pressure until solidification is completed.

Figure 3.4 shows a card laminating machine. In this system stacked foils are premounted together, whereby under and over the pre-mounted sheets thin, highly polished laminating metals plates are inserted. For example, eight pre-mounted sheets are stacked together for one batch. In the next step the batches are brought between the heating plates which are distributed over different stories. Afterwards the heating plates are pressed together and the temperature and pressure is controlled using a defined process profile.

In most cases the card layers are of thermoplastic foil material. In special cases non-thermoplastic materials which are coated with an adhesive can be laminated. Beyond this other connecting mechanisms are also conceivable as a special case. The first Eurocheque card was, for example, a paper-laminated card. The inlay of the Eurocheque card had a security paper inlay with a watermark and security string, which were laminated reciprocally with plastic foils and on the back surface a signature stripe also of security paper was carried.

Figure 3.4 Card laminating machine with automatic sheet feeder (source: Bürkle)

After lamination the single cards have to be punched out of the finished sheets. Figure 3.5 shows an automatic punching machine with an automatic sheet feeder.

Figure 3.5 Card punching machine with automatic sheet feeder (source: Mühlbauer)

In this machine three single cards from one row are punched out at the same time with three side by side placed punching tools.

In principle, the punching out of a single card works as shown in Figure 3.6. The laminated sheet is positioned above the matrix. Then the sheet is clamped between the punching matrix on the clamp. Now the punch can punch out the single cards. The size of the punch and the matrix have to be optimized for the plastic material used in order to reach the required size of the punched-out card described in the standards.

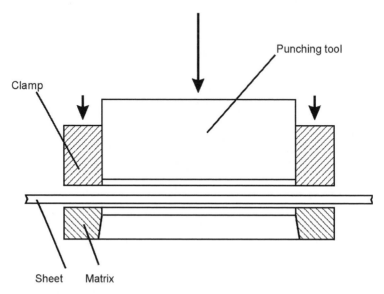

Figure 3.6 Principle of punching a single card out of a laminated sheet

The following detailed description of the laminating process is limited to laminate thermoplastic foil materials.

3.1.1   Laminating process

The usual connecting mechanism of the card layers when laminating is based on the characteristic of thermoplastic foils to be able to pass through different condition phases under supply of heat.

Figure 3.7 shows the temperature behaviour of a thermoplastic foil. As characteristic for the status of the foil, the modus of elasticity is displayed on the vertical axis and the temperature on the horizontal axis. In the temperature range T_1 the foil material is in the solid state (so-called glass status). In the temperature range T_2 the foil material is softened. Within this area there is a special point as the glass transition temperature which is defined as the glass temperature T_G. After that follows area T_3, in which the material is soft, flexible and ductile. With further heating the material melts. The transition point at this area is T_S, the melting point of the foil material. After this point the foil acts like a fluid (T_4). If the material is heated up further into phase T_5, decomposition of the material begins.

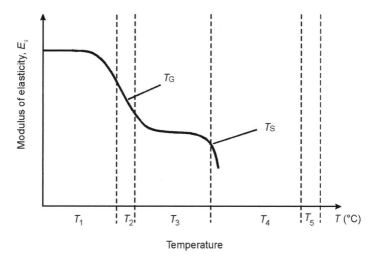

Figure 3.7 Temperature phases of a thermoplastic foil

Laminating cards from thermoplastic foils occurs in the temperature range T_3, i.e. between the glass temperature T_G and the melting temperature T_S.

Crucial for the outcome of the laminating process are on the one hand the thermoplastic characteristics of the materials which can be laminated and on the other hand the laminating parameters of temperature, time and pressure in the heating and cooling phases.

The laminating temperature depends on the softening temperature of the laminating material. The laminating time depends on the temperature transition coefficient and the thickness of all materials which are fixed between the heated plates of the laminating machine. The pressure during heating must be high enough to press the air out between the laminating layers and large enough to prevent the laminating layers from shrinking by adhesion of the laminating plates due to the heating up.

During the cooling period pressure has to be applied to the plastic foils to ensure good adhesion of the single plastic foils with each other. Also optical inhomogeneities and blisters can be prevented on the surface of the laminated foils. While cooling down the surface of the lamination plates the card adapts its final surface. This could be mat (sand blasted plates) or glossy (highly polished plates). After the complete lamination process a plastic sheet is presented with 12, 18, 24 or up to 48 cards. To obtain individual cards they have to be punched out of the plastic sheet. Afterwards the cards go through a visual inspection before additional card elements like the magnetic stripe or a chip module are applied to the card body.

Figure 3.8 shows the production flow of a three-layer card (Section 3.1.2).

Figure 3.8 Production flow of a three-layer card body using the laminating technique

3.1.2   Card structure

Depending on the use of the laminate smart card different constructions of plastic layers are in use. In the following some typical structures are briefly described.

Mono-layer card

The mono-layer card consists of only one opaque foil in the final card thickness (800 µm). In Figure 3.9 shows a mono-layer card, which was printed on both sides. The printed surface is usually covered with a transparent protective varnish.

Figure 3.9 Structure of a mono-layer card with external printing and protective varnish

The advantage of the mono-layer card as opposed to multi-layer cards involves the fact that the manufacturing steps of plastic sheet assembly and laminating are omitted.

Depending upon the application, the mono-layer card can be subjected to a laminating process for refinement of the surface, for example for the production of a highly polished surface. In this case only the process step plastic sheet assembly is required.

Two-layer card

Figure 3.10 shows a two-layer card. The card consists of two opaque foils, which are printed on their surface. The disadvantage of the two-layer card with respect to the laminated one-layer card is the additional manufacturing step necessary for the plastic sheet assembly. The advantage: the thickness of the foils is only half the mono-layer foil thickness. This is an advantage in printing, because some printing machines cannot handle 800 µm mono-layer foils. Also the printing of mono-layer foils increases the possibility of misadjustment of the design when printing on both sides of the layer. This reduces the yield. When using a two-layer card the front- and back-side printing can be separated and in the case of printing problems only half the amount of foils is wasted. Here an exact calculation is appropriate, as to what extent the increased rejects can be made up for by saving of the process step plastic sheet assembly.

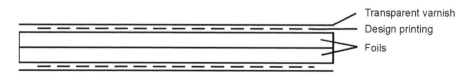

Figure 3.10 Structure of a two-layer card with external printing and protective varnish

Three-layer card

From the multiplicity of possible cards for three-layer construction here we describe two possible constructions.

Figure 3.11 shows a three-layer card. The card consists of three opaque foils. The outside foils are printed. Their thickness is selected in such a way that they can be manufactured with tolerances as low as possible.

The advantage of the three-layer card in comparison to the two-layer card is that the thickness of the foils can be chosen in such a way that the printing press works more effectively and the reject is reduced to a minimum. The disadvantage of this card is the additional expenditure in sheet assembling through the larger amount of foil positions.

For example, the thickness of the foils can be chosen as follows:

Opaque overlay foil   approx. 200 µm
Opaque inlay   approx. 400 µm
Opaque overlay foil   approx. 200 µm

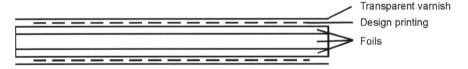

Figure 3.11 Construction of a three-layer card with external printing and protective varnish

Figure 3.12 Structure of a three-layer card with internal printing and overlay protecting foil

Figure 3.12 shows another three-layer card. This card consists of an opaque inlay which is laminated on both sides with transparent overlay foils. The opaque inlay is printed on both sides. The overlay foils are laid out as thin as possible (e.g. 60 µm). This is to prevent edge-sloping, covering printing from the possibility that at the edge of the card the overlay foil can be replaced easy. A thin overlay foil cannot be replaced easily because of less stiffness of the thin foil in comparison to a thick overlay foil. The laminating adhesive between the overlay foil and inlay foil can be optimized by the application of laminate-adhesive-coated overlay foils.

An advantage of the transparent overlay foil in comparison to the protective varnish is better scratching and abrasion resistance of the printing which is present under it.

For example, the thickness of the foils can be laid out as follows:

Transparent overlay foil   approx. 60 µm
Opaque inlay   approx. 680 µm
Transparent overlay foil   approx. 60 µm

Four-layer card

By dividing the inlay of a three-layer card into two foils a four-layer card is the result. The advantages are the same as those described for the mono-layer card in comparison to the two-layer card as well as reduction of printing waste. Obviously the expenditure of assembly of the foils increases. In the case of the structure with external printing, described previously, the following structure results:

External opaque foil   approx. 200 µm
Opaque inlay   approx. 200 µm
Opaque inlay   approx. 200 µm
External opaque foil   approx. 200 µm

Figure 3.13 shows the four-layer card.

A further advantage of this structure is that the card can be structured from four equal foils, which simplifies the purchase and stockpile of the foils.

Transparent varnish
Design printing
Foils

Figure 3.13 Structure of a four-layer card with external printing and protecting varnish

Figure 3.14 shows a further realization of the four-layer card. This card is a four-layer card with transparent overlay foils. If the printing needs to be protected with transparent overlay foils then the card construction looks like the following:

Transparent overlay foil	approx. 60 µm
Opaque inlay	approx. 340 µm
Opaque inlay	approx. 340 µm
Transparent overlay foil	approx. 60 µm

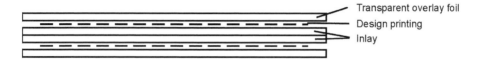

Transparent overlay foil
Design printing
Inlay

Figure 3.14 Construction of a four-layer card with protected design printing

Exchanging the inlay into two thinner layers offers the possibility of bringing in invisible security between both layers — characteristics which can be identified by special devices.

With all the described examples the foil thickness was selected in such a way that the structure of the card is symmetrical. This is important for the flatness of the card, especially if the foils have different coefficients of thermal expansion.

Further card structures

The card structures described are selected examples from the abundance of possible card constructions. Depending upon the application of cards and card items, in practice a set of different special constructions were developed and implemented. The following lists some examples:

• For an optimal laser inscription or laser photo engraving of the card, the use of doped foils in thickness gradation is necessary
• The implementation of a multiple laser image (MLI) required for optimal laser engraving also the application of several differently doped foils, those to be laminated one above the other
• For certain visual card items it is often necessary to make a special card construction. For example, with some card applications several overlay foils with different levels are printed in such a way that after laminating the foils the impression of a spherical picture

develops. Frequently the security element has to be in certain layers of the card construction

3.2 Injection moulding

The proven mass production procedure of injection moulding offers itself for the production of card bodies for smart cards. The substantial advantage of this procedure is the fact that the cavity for later embedding of the chip module is produced during the injection moulding of the card.

The raw material used for the injection moulding is plastic granulates, which melt at temperatures between 200 and 400 °C. They are pressed with a pressure into pre-heated injection moulds between 800 and 2300 bar. The injection mould is sealed for the melted plastic material; however, air can escape when the material penetrates into the mould. The tool temperature favours flowing of the material into all corners and edges of the form, causing — after complete filling out of the form — the immediately solidification of the melt within the cycle time. Afterwards the form is opened and the work piece is removed or ejected. Very often a rework of the work piece is necessary, for example removing the runner.

In Figure 3.15 an injection-moulded card with the chip module cavity is shown schematically.

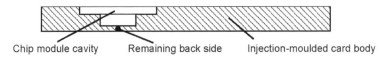

Chip module cavity   Remaining back side   Injection-moulded card body

Figure 3.15 Cross section of an injection-moulded card body

A special difficulty during the production of such cards is the relatively thin remaining material at the back side of the cavity. It has a thickness of approximately 170 µm, whereby the thickness of the card amounts to approximately 800 µm. Sometimes in this thin area the material breaks or the flatness of the surface is not guaranteed. A solution to this problem is the following procedure which works well in practice: As schematically shown in Figure 3.16, the injecting channel is attached close to the cavity on the short edge of the card. In the tool a movable stamp is arranged, which is pressed into the form after filling the form and before the material solidifies. This produces the cavity and the thin rear wall.

One of the most important features of an injection-moulded card is the accuracy of size. But shrinkage is crucial with injection moulding (e.g. the proportional deviation of the mass of the work piece from the masses of the tool). Additionally, the shrinkage also depends on the chosen material and process parameters.

The injection-moulded card is normally printed in special single-card printing machines before the embedding of the chip module. The printed surface is over-printed with a transparent protecting varnish. Here a further important demand for the surfaces of the card is the printability of the injection-moulded card.

Further the adhesive surface for the chip module must be so arranged that a good adhesive of the chip module is possible. Depending upon the chosen bonding technique a special condition of this surface is called for. In all cases all surfaces must be free from dirt and parting agents.

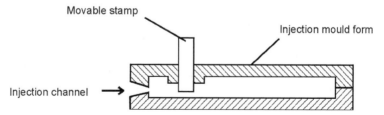

Figure 3.16 Pattern of the injection-moulding technique for smart cards with movable stamp to create the cavity into the card body

Beyond the here described solution of a moulded card body with a cavity for the later embedding of a chip module there are some advanced developments already realized. One of these solutions is the so-called in-mould technique.

Here a special type of chip module is placed and fixed in the mould form before the injection mould material is pressed into the form. After solidification a complete smart card is produced.

It is also possible to insert some pre-printed layers into the moulding form. During the injection of the mould material the layers are pressed onto the surface of the card and the card body is created including the final design. This technique is already used successfully in other industrial areas.

The plastic material acryl butadiene styrene (ABS) has become generally accepted for the production of injection-moulded cards. ABS is a very good material for the injecting technique and it is free from chlorine and hence more environmentally friendly than PVC.

3.3 Paper card

Section 2.2 described how the card body can also be made of paper or cardboard. This offers itself in particular for cards which are mechanically not strongly stressed and not exposed to humidity in use. Particularly cards with limited lifetime, e.g. disposable cards, can be considered for the application of paper cards. Disposable cards are, for example, tickets for sports events, single tickets or daily cards for public transport, calling cards with a small amount of units and cards used for similar applications.

Figure 3.17 shows an example of a paper card.

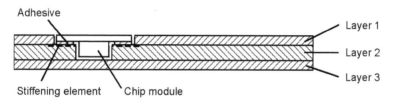

Figure 3.17 Possible structure of a paper card with chip module

This card consists of three cardboard layers. Layer **1** is suitably punched for the size of the chip module carrier tape and layer **2** is suitably free-punched for the size of the chip module glob top.

With the embedding of a chip module into the paper card the connection of the chip module to the paper surface can be manufactured optimally with an appropriate adhesive. The durability of this connection is however problematic because the small internal adhesive of the paper can be easily removed by splitting the paper itself. A solution for this problem, for example, exists in stiffening the paper in the area of the chip module bonding.

In Figure 3.17 a stiffening of the chip module bonding area is suggested. The stiffening can be achieved, for example, by locally gluing a piece of textile within this area. In addition, it is possible that the entire layer two consists of strengthened cardboard. Layer **3** forms the cards back side and the rear wall. Layers **1** and **3** can already be pre-printed.

The production of the smart card can be undertaken with both the production techniques of mounting (see Section 4.2) and laminating. During production using the mounting technique, the three-layer sheets are positioned towards each other and glued together. Afterwards, individual card bodies are punched out of the sheet. In a further process, step chip modules are embedded in the cavity of the paper card.

Production using the laminating technique integrates the implanting of the chip module into the production process of the card body.

For production of the cardboard card various production processes from the packaging industry can be used, which mostly require very short cycle times. Various gluing procedures are also possible with dispersion adhesive or with glue. Also hot melting adhesives can be used, because of the high thermal stability of cardboard at high temperature, when short cycle time is processed.

4 Embedding of chip modules into smart cards

In the preceding chapter the production of card bodies was described. In this chapter different techniques are introduced for the embedding of chip modules into card bodies and an overview of different possibilities for smart card production is given. The following smart card manufacturing techniques are used:

4.1 Embedding by lamination technology

Here a finishing technique for smart cards is aimed for, with which the chip module is built into a card layer before laminating the complete card. In Figure 4.1 a three-layer card with external printing is shown. The core foil has for instance the same thickness as the glob top of the chip module. The overlay foil on top of the inlay also has a punched window but in this case the window is the size of the contact tape of the chip module. The thickness of the overlay foil has the same thickness as the contact tape. The upper overlay foil is likewise provided with a punching window. This is the size of the contact plate of the chip module.

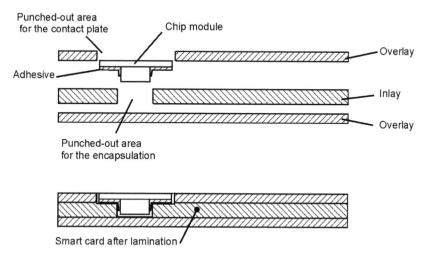

Figure 4.1 Principle way for producing a smart card using the laminating technique

Before the chip module is punched out of the chip module carrier tape (normally 35 mm width tape with two modules side by side), heat-activatable adhesive is applied on to the back side of the chip module carrier tape. After the chip module is punched out of the carrier tape the chip module is pre-mounted in the cavity of the inlay. For better handling of the chip module and inlay the heat-activatable adhesive is activated with a heated stamp which bonds the chip module to the inlay. This equipped inlay is assembled with the punched-out overlay foil and the back side overlay foil. The foil package then undergoes the laminating process. After the laminating process, the single card has to be punched out of the sheet. The punching has to be done in a very precise manner as it influences the later position of the electrical contacts of the chip module which are described in the standards.

Advantages:
- The manufacturing machines, which are necessary for the laminating technique are simple machines that can be usually taken over from other applications. Punching out the plastic sheets takes place in efficient punching machines. For the assembly of sheets with chip modules the well-known technique of SMT (surface mounting technology) can be used
- As the chip module is connected to the card during the laminating process the plastic totally encloses the chip module

Disadvantages:
- The biggest disadvantage of the laminating technique is the amount of rejects. With each rejected card from the laminating process a chip module is also thrown away, which is very expensive in terms of microcontroller chips. Rejects from the laminating process could be due to :
  - misadjustment of the image
  - punching tolerances to high
  - scratches and dirt on the surface of the card
  - wrong process parameter which can destroy or damage the chip

  Therefore the laminating technique can only be used under the following conditions:
  - for economical chip modules with cheap small chips
  - for simple cards with only a few additional card elements
  - where the laminating process runs stable and with a high yield

- Different thermal expansion quotients of the card materials on the one hand and the chip module materials on the other hand for the laminated card lead to permanent deformation of the back side within the chip module area. If chip modules with small glob top are used the deformation should be acceptable

4.2 Embedding by mounting technology

Mounting techniques are those smart card production techniques, which are based on the fact that a cavity is pre-produced in the card body before — in the following production step — a chip module is insert into this cavity.

In the preceding chapter production of card bodies using the laminating technique or injection-moulding technique was described. While the injection-moulding card is already manufactured with a chip module cavity, the chip module cavity must be produced in a separate production step for a laminated card. That is usually done via milling.

4.2.1 Dimensions of the cavity

Figure 4.2 shows schematically the substantial items from which the smart card is assembled.

For a better overview only the chip module area of the smart card can be seen in the figure. As evident from the figure, the substantial items of a smart card are the chip module cavity, the chip module and the adhesive for the connection of the chip module with the card body, and the card body itself.

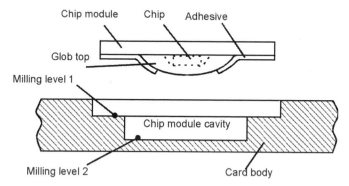

Figure 4.2 Elements of a smart card (chip module area of the smart card is enlarged)

The basic problem for the smart card is the fact that the chip is stiff and brittle. On the other hand the smart card, in which the chips are embedded, has to be very flexible to fulfil the demands of daily usage. One way to protect the chip is to ensure the chip is placed in the centre of a carrier tape encapsulated with a glob top to give it additional stiffening. A further solution to the problem is flexible suspension of the chip module in the cavity. As shown in Figure 4.2, the chip module cavity consists of two levels.

On level **1,** the chip module is connected through the card body using a flexible adhesive. The used adhesive is free punched in the area of the glob top in such a way that there is a small overlap of the adhesive over the glob top, which also brings some more stabilization to the chip module.

The cavity size of level **2** is chosen in such a way that the glob top does not touch the bottom of the cavity. Now the chip is almost hanging free in the cavity and released from the bending loads. Figure 4.3 shows the chip module embedded in the cavity.

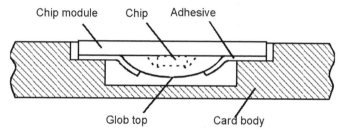

Figure 4.3 Embedded chip module in the milled card cavity

In conclusion many influences have to be considered for good performance of a smart card, e.g. the geometrical conditions of the chip module cavity, the characteristics of the adhesive, the adhesive surfaces and later on the process parameters of the embedding.

In the following the influence of the geometry of the chip module cavity is considered. As an introduction to the topic of geometrical conditions of chip module embedding, first two different embedding principles are described.

The difference between both procedures exists in the different definition of the reference level when milling the cavity and the resulting different tolerance calculation.

Case I: Reference level on the back side of the card body

Figure 4.4 shows a card with chip module cavity, where the reference level for milling the cavity is on the back side of the card. The card is clamped on the machine table and processed. The height of the milling tool is adjusted relative to the level of the machine table. The result of the milling process is the card shown in Figure 4.5.

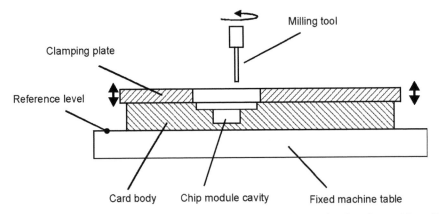

Figure 4.4 Clamped smart card in the milling station with the reference level on the machine table

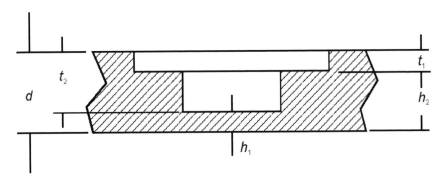

Figure 4.5 Milled cavity with reference on the back side of the card

When the reference level is the back side of the card the tolerances of the depth of the upper cavity t_1 are equal to the total of the tolerances of the card thickness d and the milling tolerances of the height of h_2. Additionally the tolerances of the rear wall thickness h_1 are much smaller, about the size of the milling tolerances. This means that the tolerances of the card thickness d enter completely into the depth t_1.

Case II: Reference level on the card surface

In Figure 4.6 the milling of the card cavity is shown with the card surface as reference level. The card is clamped and processed with a machine table which is under the card and the clamping plate is situated over the card.

With this type of card milling the tolerances of the depth of the upper cavity t_1 are equal to the milling tolerances, while the tolerances of the rear wall thickness h_1 are equal to the total tolerances of the card thickness and the milling tolerances of the depth t_2. This means that in this case the tolerances of the card thickness d enter completely into the tolerances of the rear wall thickness h_1.

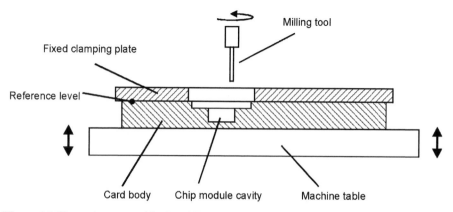

Figure 4.6 Clamped smart card in the milling station with the reference level on the card surface

The implementation of the cavity milling station, as shown in Figure 4.6, clamping the card from the back side, is constructional not easy to realize in mass production. This problem is solved by the fact that the same machine is used as shown in Figure 4.4. The card thickness d is measured before milling. In dependency of the thickness d, the milling depths t_1 and t_2 for each card are calculated and adjusted.

Depth conditions of the chip module cavity and their reference points as well as the rear wall thickness of the chip module cavity are important parameters for the production and function of the smart card. In the following these parameters and their effects are explained. The study is made on case **I** where the cavity with the rear side was the reference level as shown in Figure 4.7.

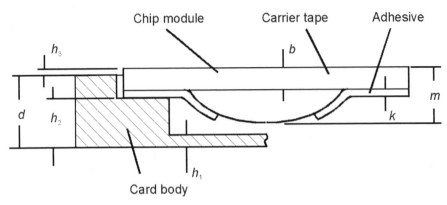

Figure 4.7 Sketch of important dimensions for the calculation of cavity tolerances

For good production and later functionality of the smart card the following conditions must be fulfilled in order that during the embedding process of the chip module into the cavity no pressure is transferred on the chip.

$$m + \Delta m - b + \Delta b - k + \Delta k \leq h_2 - \Delta h_2 - h_1 - \Delta h_1 \qquad \text{(a)}$$

With each dimension the tolerances are added with plus and minus:

$m \pm \Delta m$	thickness of the chip module
$b \pm \Delta b$	thickness of the carrier tape
$k \pm \Delta k$	remaining thickness of adhesive after embedding(*)
$h_1 \pm \Delta h_1$	milling depth 1
$h_2 \pm \Delta h_2$	milling depth 2 (normally Δh_1 and Δh_2 have the same value)

(*) Remark: It should be noted that the thickness and the thickness tolerances of the adhesive are usually larger in unfinished status than the remaining thickness after embedding. The remaining thickness k and the tolerance thickness Δk are to be determined by trials and measurements.
The total thickness of the smart card is limited according to the standard to a maximum of 840 μm. The consequence is:

$$h_2 + \Delta h_2 + k + \Delta k + b + \Delta b \leq 840 \ \mu m \qquad \text{(b)}$$

$b \pm \Delta b$	thickness of the carrier tape
$k \pm \Delta k$	remaining thickness of adhesive after embedding(*)
$h_1 \pm \Delta h_1$	milling depth 1

If relationship (b) is fully exhausted, then the equation changes to:

$$h_2 + \Delta h_2 + k + \Delta k + b + \Delta b = 840 \ \mu m$$

For the protrusion of the chip module above the surface of the smart cards h₃ is:

$$h_{3\,max} = h_2 + \Delta h_2 + k + \Delta k + b + \Delta b - d + \Delta d \qquad\qquad \text{(c)}$$

$h_{3\,max}$	maximum protrusion of chip module/card body
$d \pm \Delta d$	thickness of the card body

For the production of smart cards the thickness tolerance of the smart card is usually limited to $800 \pm 40\ \mu m$.
Under this condition $h_{3\,max}$ is:

$$h_{3\,max} = 840\ \mu m - 800\ \mu m + 40\ \mu m = 80\ \mu m$$

This means that, with a small probability if all relevant dimensions are present at their maximum, a protrusion of 80 µm can occur. Until recently this value of protrusion was admissible according to the standard. It led, however, to problems when stacking and sorting the cards one above the other, and with badly punched chip modules burr scratching of the card above. In the newest revision of the standards the protrusion h_3 was limited to + 50 µm, measured from the top of the smart card surface. Not only is the protrusion described in the standard, but also a value is given as to how deep the chip module could be placed under the level of the smart card surface, this value is –100 µm.

Milling geometry is calculated from Equation (a). The most important dimension is the chip module thickness m. Today available chip modules have a maximum thickness of 580 µm. Various developments, above all the possibility of production of thinner chips, will also enable the realization of thinner chip modules in the future. In the following a sample calculation with assumed mass and tolerances demonstrates the limits and complexity of the milling process.

Assume (all dimensions in µm):

$m \pm \Delta m$	$= 560 \pm 20$	thickness of the chip module
$b \pm \Delta b$	$= 190 \pm 30$	thickness of the carrier tape
$k \pm \Delta k$	$= 30 \pm 5$	remaining thickness of adhesive after embedding
$\pm \Delta h_1$	$= \pm \Delta h_2 = \pm 15$	milling tolerances
$h_{3\,max}$	$= 50$	maximum protrusion of chip module/card

Calculation of h_2 out of Equation (c):

$$h_2 = h_{3\,max} - \Delta h_2 - k - \Delta k - b - \Delta b + d - \Delta d$$

$$h_2 = 50 - 15 - 30 - 5 - 190 - 30 + 800 - 40 = 540\ \mu m$$

Calculation of h_1 out of Equation (a):

$$h_1 = h_2 - \Delta h_2 - m - \Delta m + b - \Delta b + k - \Delta k - \Delta h_1$$

$$h_1 = 540 - 15 - 15 + 30 - 5 + 190 - 30 - 580 = 115\ \mu m$$

As shown in the above calculation the remaining rear thickness h_1 in case **I** does not depend on the card thickness and indicates only small tolerances, i.e. the milling tolerances. The tolerances of the card thickness reflect themselves in the maximum protrusion of the chip module/card.

If one sets up the relations for case **II** similarly to the relations specified above, one will detect that in case **II** smaller ranges of tolerance enter the protrusion of the chip module/card, while the entire thickness tolerances of the card affect the rear wall thickness h_1.

In a general manner the calculations specified above can be also used for the injection-moulded card. With the injection-moulded card the tolerances of the card thickness are smaller, which leads to the fact that for h_1 larger values can be used than with the milled card. That is particularly important for the injection-moulded card, since the rear wall thickness determines the flow behaviour of the material and thus the production cycle time and the production yield.

The arithmetically calculated rear wall thickness h_1 is very small. This has the consequence that the rear wall deforms heavily and breaks during dynamic load. For this reason in both the milled card and the injection-moulded card it is aimed to increase the rear wall thickness. A thicker rear wall means better protection for the chip module and for the card, higher lifespan and better appearance of the card with smaller deformation of the card surface within the rear wall area.

In order to achieve this requirement, one tries to reduce chip module thickness, and material and processing tolerances.

A further possibility for increasing the rear wall thickness is not to fulfil the complete span of Equation (a). The following considerations can be made valid:

- A prerequisite for Equation (a) is that even if all tolerance limits are present the chip module should never touch the rear side of the cavity because during embedding the chip module and in particular the chip, could be damaged. To avoid chip damage there are some possibilities for protecting the chip. One way is right design of the stamp which presses the chip module into the cavity during the embedding process. The stamp only needs to have a window in the area where the chip is, so that no pressure is loaded on the chip. In addition, the chip module can be structured in such a way that the brought in forces will be caught within the chip module and transferred to the chip
- Further the probability for the fact that all tolerances even in the unfavourably direction for Equation (a) achieve their extremum, is infinitesimally small. Therefore a statistical calculation of the rear wall thickness is recommended.

Statistical calculation of the rear wall thickness

If one presupposes a normal distribution for all relevant tolerances, then the statistical rules for normal distribution can be made use of. In Figure 4.8 the normal distribution for the value X is shown. The average value X_m occurs with largest probability, the maximum and minimum value of X_{min} and X_{max} will occur with smallest probability.

A reduced range of tolerance T (difference to $X_{max}-X_{min}$) is defined which shows six times the size of the standard deviation of the value of X. This value for T is selected as simple mathematical relations for tolerance summation can be set up for the calculation.

For this selected value, T gives from the curve of the normal distribution a probability in the area of $2 * 0.135 = 0.27$ % for exceeding or falling below the reduced range of tolerance.

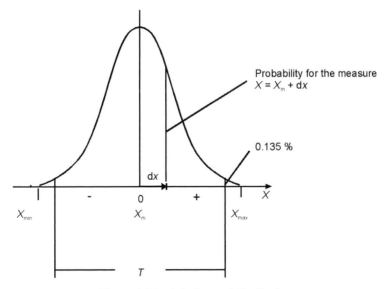

Probability for the measure
$X = X_m + dx$

0.135 %

dx

0

\-   +

X_{min} X_m X_{max}

T

Figure 4.8 Statistical normal distribution

If T is set as a range of tolerance, then the following relationship applies with a sum tolerance T which results from several tolerances T_1, T_2, T_3...:

$$\boxed{T^2 = T_1^2 + T_2^2 + T_3^2 + ... + T_n^2}$$ **(d)**

This formula reflects the fact that the more tolerances T_1, T_2, T_3... add themselves to a sum tolerance T, the value of the sum tolerance T rises. However, it does not rise like the arithmetic total of all tolerances as the more tolerances T_1, T_2, T_3... are involved towards the sum tolerance the probability reduces that all of them occur with their unfavourable extremum.

In order to be able to calculate the rear wall thickness h_1 from relationship (d) a new parameter s has to be introduced. It defines the distance of the deepest point of the chip module to the rear wall surface as shown in Figure 4.9.

$$\boxed{s = h_2 + k + b - m - h_1}$$ **(e)**

It is to be taken from relationship (e) that T_s (tolerance range of s) depends on the tolerances T_{h2}, T_k, T_b, T_m and T_{h1}.

$$T_s^2 = T_{h2}^2 + T_k^2 + T_b^2 + T_m^2 + T_{h1}^2$$ **(f)**

Hereby is T_X the total tolerance area:

$$\boxed{T_X = X_{max} - X_{min} = 2 * |\Delta X|}$$ **(g)**

If we take the same dimensions from the last sample from Equation (g) we get:

$T_s^2 = 30^2 + 10^2 + 60^2 + 40^2 + 30^2 = 7100$
$T_s \approx 84 \; \mu m$

To avoid the chip module touching the rear wall of the cavity the average of s is taken:

$s = T_s / 2 = 84/2 = 42 \ \mu m \approx 40 \ \mu m$

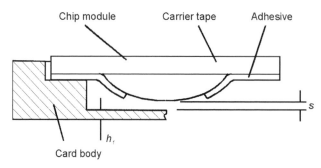

Figure 4.9 Definition of the distance s for the statistical calculation of the remaining rear wall thickness

Now the value for the remaining rear wall thickness h_1 can be calculated with a worked out value for s (approximately 40 μm). Equation (e) can be taken for the calculation:

$h_1 = h_2 + k + b - m - s$

Here no more tolerances occur. All tolerances are already considered with the calculation of s. The above arithmetically calculated value is used as the value for h_2.

$h_1 = 540 + 30 + 190 - 560 - 40 = 160 \ \mu m$

Now all relevant dimensions for the depth (h_1, h_2) are calculated.

4.2.2   Gluing technology

The German DIN 8593 standard describes in a good way how multifarious the joining of parts can be. Figure 4.10 gives a mixture of the norm. We will only concentrate on smart cards in section 4.8 of the standard.

Additionally the types of adhesive can also be divided depending on raw material (epoxy resin, natural rubber and glue), according to the type of processing (fusion adhesive, heat-activatable adhesive and contact adhesive), used processing temperature (cold adhesive and hot adhesive), form of delivery (adhesive film and sticking pin), applications (paper adhesive, wood adhesive or PVC adhesive), nature of the surface or on the basis of other criteria. The expressions, which are used in the following, are common in the smart card world and do not necessarily correspond to a systematic designation.

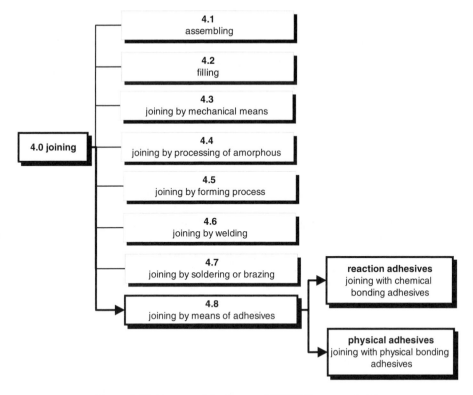

Figure 4.10 Abstract of the German DIN 8593 paragraph 4.0

Physics of joining

The materials of which adhesives consist of are so diverse that only one common predicate for all adhesives can be made — that they consist of non-metallic materials. Even this predicate can be retracted if one considers electrical conductive adhesives, which contain metal particles. They do not contribute to sticking, but to the electrical conductivity.

In order to understand the mode of operation of an adhesive the most important terms of the adhesion method are explained in the following.

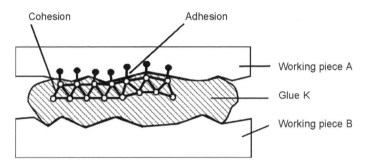

Figure 4.11 Mode of operation of bonding

Figure 4.11 schematically shown bonding between two work pieces.

The two work pieces **A** and **B** are bonded together with adhesive **K**. The molecules of the adhesive have such structures that they form long chains after termination of the bonding procedure. Additionally, they have started from the boundary surface of the adhesive bound with molecular attraction, which can be used for the linkage with the surface of the work pieces **A** and **B**.

The strength of the bonding depends on the following mechanisms:

Coherence

Coherence is the name given to the internal forces of the adhesive. The coherence is responsible in preventing the adhesive from being removed by cracks.

The coherence is a characteristic of the adhesive itself. The processing parameters must be adapted to be optimal to the adhesive in order that the maximum possible coherence of the adhesive can be achieved. This means that with the failure analysis of a broken bond the cause of the bonding failure can be found in the coherence, if the adhesive were split into two sections with the adhesive remaining completely at the two materials. In this case it has to be verified if the working procedures of the adhesive manufacturer were followed, otherwise another adhesive has to be used which is able to fulfil the request of the demands.

Adhesion

Adhesion is the name for the forces which connect the work pieces surface with the adhesive surface. They prevent the adhesive from becoming detached from the work pieces surface.

Adhesion depends on the type of adhesive, the work piece materials and on the condition of the material surface.

With failure analysis, attention has to be put on the adhesion, adhesive may be lost partly or completely from the work piece surface.

For good connection with the material the following mechanisms are crucial:

Wetting

An optimal bonding takes place among other things by the fact that the adhesive comes into direct contact with the entire surface of the work piece, thus the adhesive strength becomes effective within its range. A prerequisite for the optimal wetting of the surface of the work piece is that the adhesive adapts to the roughness depth of the work piece and penetrates into the depths of the surface texture. Subsequently the adhesive surface is larger and also the bonding strength is larger.

Surface energy

Figure 4.12 schematically shows the effect of surface energy. A higher surface energy means a better wetting of the surface. This characteristic is also used for the measurement of surface energy, as liquid drops with well-known wetting characteristics are dispensed on the work piece surface. The angle between the baseline of the drop and the tangent at the drop boundary is measured. This measurement method is called the sessile drop method.

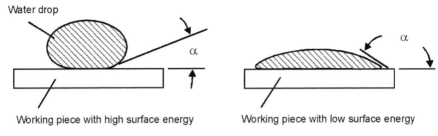

Figure 4.12 Sessile drop method for measuring the surface energy of a work piece

By pre-treatment of the material surface it is possible to improve the conditions for good wetting and a higher bonding strength. A very practical method for surface pre-treatment is plasma treatment. The work piece is brought into a chamber, which is evacuated and afterwards filled with a process gas. By stimulating the process gas with microwaves in the direct area of the surface a plasma reaction is produced which removes organic contamination from the surface and increases the surface energy.

Roughness

Only few adhesives have a better bonding result if the surface of the work piece is smooth. For most adhesives one obtains better results by roughening the material surface with optimal wetting of the surface at the same time.

The reason for this is that with a rougher surface the total surface of the work piece increases and therefore the total strength of the adhesive increases too.

A further reason is that roughening up the surface profile results in many small adhesive surfaces, which are sloping in relation to the bonding surface. The consequence is that the adhesive strength on these surfaces in different directions becomes effective and so this bonding can resist dynamic loads with variable load directions in a better way. In the case of graining by material removal (for example, sandblasting) partial pits in the surface result. If the adhesive can moisten this surface well, the adhesive hooks itself in these pits, and tiny closing shapes result, which increase the bonding strength. Figure 4.13 schematically represents these effects.

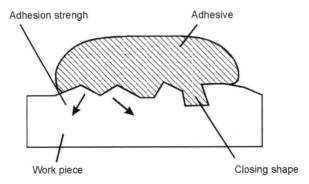

Figure 4.13 Influence of the surface roughness on bonding

Types of adhesives

As already mentioned at the beginning of this section we use the following types of adhesives which are common in the smart card world. Usually the naming of the types of adhesives is according to the processing type of the adhesive.

For the production of smart cards the following adhesives are used:

- Heat-activatable adhesives
- Double-faced adhesive tape
- Compound adhesives
- Thermoplastic adhesives
- Liquid adhesives

Heat-activatable adhesives

Heat-activatable adhesives are available as tape in rolls. The tape roll contains a protection tape, which usually consists of silicone-coated paper. This protection tape is used in the production of the adhesive as a carrier for the adhesive, and remains adhered easily on the adhesive and is cut and supplied with it.

Heat-activatable adhesives can consist among other things of the following materials:

- Natural rubber
- Nitrile rubber
- Acrylate resin
- Polyurethane

Request:
The request for the adhesive is given from the smart card production:

During the production of a smart card the adhesive is first pre-laminated on the chip module carrier tape and is punched out with the chip module later in the production process. The adhesive-laminated chip module is inserted into the card cavity, activated with a heating stamp and in the next step fixed with a cooling stamp.

- Mechanical characteristics (thickness, thickness tolerances, width, total length per roll and roll dimensions)
- Free from dirt and dust
- No health damage during the processing of the adhesive
- The smart card should not pollute the environment during daily usage
- Demand for the chip module surface (material, roughness depth and pre-treatment)
- Demand for the card material
- The adhesive should be as suitable as possible for all card materials
- Laminating parameters (temperature, time, pressure)
- After pre-laminating the silicone paper should be able to be easily removed from the adhesive
- Punishing capability of adhesive with silicone paper and laminated adhesive on the chip module tape
- Embedding parameters (temperature, time, pressure)
- Thickness tolerances after the embedding
- Behaviour of the embedded smart cards in a batch and during shipment

- Binding power, dynamic strength, strength during temperature and humidity influence, resistance against UV radiation

Advantages:

The advantages mentioned result in the case of an optimal combination of bonding characteristics and processing parameters:

- In processing, heat-activatable adhesives offer clear advantages especially compared to liquid adhesives:
    - easy storage conditions
    - the remaining adhesive can be used again after manufacturing stops and longer manufacturing interruptions
    - pre-laminated chip modules can be stored temporarily
    - the machine units for the processing of heat-activatable adhesives are simple and space-saving
    - there is usually no odour nuisance and noxious gas development during the processing
- The cycle time for the processing of heat-activatable adhesives can be very short. Many adhesives require an activation time, under one second. Additionally the activation can be executed in several separated heating stations. It is possible to reduce the process time per station to fractions of a second
- The heat-activatable adhesive permits an optimal process stability. The wetting of the adhesive surface arises as a result of the punching outline of the adhesive and the embedding pressure. By using a suitable work piece surface (roughened up, fat- and dirt-free) with suitable material and optimal process parameters, a good bonding result over the entire adhesive surface is guaranteed
- In the finished smart card the heat-activatable adhesive offers a high bonding strength and an optimal long life stability. The cause for this is the fact that the heat-activatable adhesive is not brittle when the adhesive is hardened as it still has a certain flexibility

Disadvantage:

- The process temperature during embedding can lead to a deformation of the card back side and to the damage of the chip, hence the materials and process parameters have to be selected optimally

Double-faced adhesive tape

There are two different forms of deliveries for the double-faced adhesive tape. One form of delivery is with carrier tape so-called 'liner' and the other one is 'linerless'. The contact adhesive with carrier tape consists of a carrier tape which is coated with an adhesive mass on both sides. Double-faced adhesive tapes without liner consist of an adhesive mass which is manufactured as film on a silicone tape and in some cases covered with another silicon tape. If the delivery form is on rolls then only one silicone tape could be used if the adhesive on the outside of the silicon tape separates easier than on the inside. This is shown in Figure 4.14.

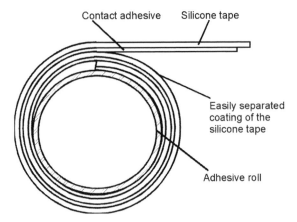

Contact adhesive   Silicone tape

Easily separated
coating of the
silicone tape

Adhesive roll

Figure 4.14 Double-faced adhesive tape on a roll with silicone tape as protective tape

In the processing of the contact adhesive special precautions are necessary for punching and removing the silicone tape. Thereby a problem is lubrication of the punching tool and adhesion of the laminated chip module tape with contact adhesive on the machine guidance or other tools which contact with the adhesive. There are different suggestions to solve this problem, which are discussed next.

Figure 4.15 shows a possibility for solving this problem. The contact adhesive is covered on both sides with a protecting silicone tape which has a free punched window in the size of the chip module encapsulation. Then one silicone tape is removed and the adhesive side is pre-laminated onto the chip module tape. Subsequently, the chip module as well as the adhesive tape and the protective silicone tape are punched out. By using special mechanics (e.g. with a vacuum) the protective silicone tape is removed from the chip module and then the chip module is embedded into the cavity of the smart card.

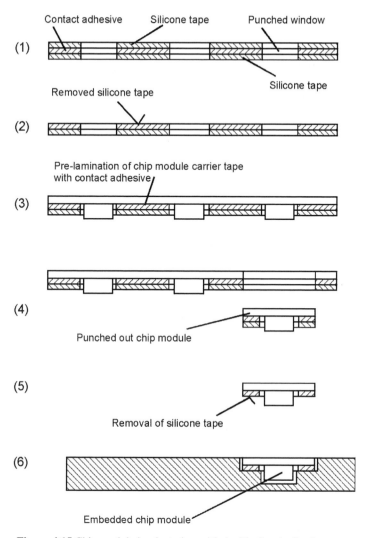

Figure 4.15 Chip module implantation with double-faced adhesive tapes

A further solution is shown in Figure 4.16. Here pre-punched contact adhesive labels are prepared on a protective silicone tape. First, a contact adhesive label is bonded into the card cavity. Subsequently, a punched-out chip module is embedded on the adhesive.

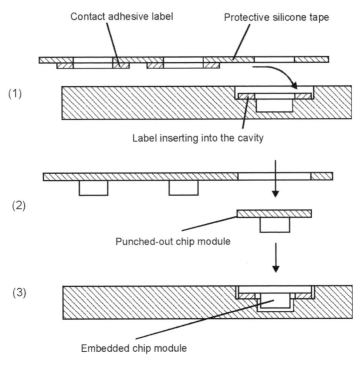

Figure 4.16 Chip module embedding with contact adhesives label

Advantages:
- All the advantages specified for the heat-activatable adhesive are also valid for the contact adhesive. Only handling of the contact adhesive is sometimes difficult with embedding
- In the processing of the contact adhesive the embedding production step itself is not critical for the process time, but punching and removing the protecting silicone tape. With comparable machine equipment this station can require a longer cycle time than with heat-activatable adhesive
- A substantial advantage of the contact adhesive is the fact that there it is no process temperature required for the embedding process, and the card body will not be deformed
- The contact adhesive remains flexible in the card. Thus the loads are caught and reduced before being transferred to the chip module

Disadvantages:
- A disadvantage of the contact adhesive involves increased machine cycle time with the problems of punching and removing the protecting silicone tape
- The adhesive strength of the contact adhesive is usually less than with heat-activatable adhesives. Chip modules for contact adhesive must normally have a larger adhesive surface and are required to have a good bonding to the smart card

Compound adhesives

With compound adhesives an adhesive is defined as consisting of several different adhesive layers. The use of compound adhesives can be due to different reasons, e.g. each adhesive layer for the respective adhesive surface can be optimized.

Adhesive layer 1

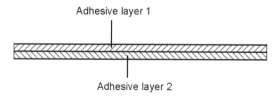

Adhesive layer 2

Figure 4.17 Structure of a two-layer compound adhesive

Figure 4.17 shows the schematic structure of a two-layer compound adhesive.

Advantages:

- As previously mentioned, it is possible by application of a compound adhesive to adapt the single adhesive layer to the two very different bonding surfaces of the chip module and the card material
- Further it is possible by the application of particularly arranged compound adhesives, for example, of a three-layer adhesive with a flexible carrier in the centre, to compensate the mechanical loads and relieve the chip module. See Figure 4.18 for this

Adhesive layer 1   Carrier tape

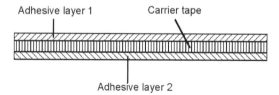

Adhesive layer 2

Figure 4.18 Structure of a three-layer compound adhesive with carrier tape to compensate mechanical loads

- By usage of a thick carrier tape it is further possible to increase the total thickness of the adhesive and it is then possible to shift the bonding area into a load-neutral level of low load of the card

Disadvantages:

- Material costs for compound adhesives are usually much higher than for simple adhesives
- With compound adhesives with increased thickness the diameter of the adhesive roll is larger or the adhesive length per roll is less, this causes more machine stops for changing the adhesive roll

Thermoplastic adhesives

Figure 4.19 schematically shows a system for the dosage of thermoplastic adhesives. The thermoplastic adhesive is present in bar form and inserted into a container which is heated from the bottom side. The adhesive melts only in the lower section, while the rest still remains in the solid state. The melted adhesive flows through a heated drain into the heated dosing needle.

The subsequent treatment of the proportioned thermoplastic adhesive is different. Here two possibilities for further processing are listed:

- Immediately after the dosage of the melted thermoplastic adhesive the chip module is embedded into the still-liquid thermoplastic adhesive and afterwards solidified with a cooling stamp
- The proportioned thermoplastic adhesive solidifies prior to the chip module being embedded. When embedding the chip module with a heating stamp the adhesive is activated again and then fixed under the pressure of a cooling stamp. In this case the proportioned thermoplastic adhesive is treated similar to a heat-activatable adhesive

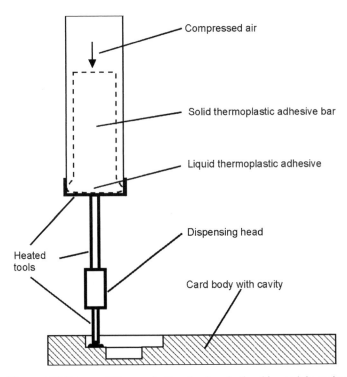

Figure 4.19 Dosage of thermoplastic adhesives in the chip module cavity

Advantage:
- If the adhesive and the embedding process parameter are chosen in the right way then cheap production with sufficient quality is possible

Disadvantages:

- Sometimes wetting of the entire adhesive surface represents a problem which has to be solved by mechanical engineering
- The adhesive outflow from the cavity onto the card surface is a problem, this could be solved by exact dosage and by special design of the adhesive surface geometry
- During dispensing of the melted adhesive the card body is thermally stressed and sometimes deformed

Liquid adhesives

There are many different types of liquid adhesive; some of them are listed in the following:

- **Cyanoacrylate adhesives**

Cyanoacrylates harden out a few seconds after the assembly of the chip module. The adhesive needs humidity at the surface of the work pieces for hardening. Therefore a relative air humidity of 40 — 60 % must be available for this process. This air humidity forms a thin humidity film which is sufficient for hardening the cyanoacrylate on the cavity surface. The hardened cyanoacrylate adhesive layer film is very thin and very stiff, but not flexible. A prerequisite for the fast and complete hardening of the cyanoacrylate — beside the humidity film — is a good fit of the two bonding surfaces (card body/chip module carrier tape) and a smooth surface on both parts

- **Silicone adhesive**

Also needs humidity for hardening as for cyanoacrylate adhesives; however, forms a flexible adhesive layer

- **Solvent adhesive**

With these adhesives the adhesive is resolved in solvents. They harden out as the solvent escapes from the bonding slit

- **Anaerobic adhesives**

Anaerobic adhesives harden by deoxidation. The processing is more favourable than with cyanoacrylate adhesives because after dispensing the adhesive must not be joined immediately. The hardening begins after embedding. By embedding the chip module into the cavity with the dispensed adhesive layer the air is pressed out and the hardening can be started.

Anaerobic adhesives also need contact with a so-called active metal (steel, brass or copper) or the addition of an activator for the hardening

- **Ultraviolet (UV) adhesives**

UV adhesives harden out under the influence of ultraviolet light. For the hardening, UV light of a particular intensity and wavelength must be chosen for every different type of adhesive layer. Since UV permeability cannot always be ensured for the smart card, a catalyst must be added to the adhesive. With this catalyst it is possible to initiate the hardening by precipitation with intensive irradiation through the bonding gap with a UV lamp. The catalyst ensures that the hardening by precipitation in the form of a chain reaction moves on into the darkest corners

- **Two-component adhesives**

As like the name implies, here two different materials must be mixed and dispensed. The hardening takes place via chemical reaction of both materials. The two materials are epoxy adhesive and polyurethane. They usually require an accurate mixing proportion. After mixing they must be processed within a given time.

There are also two-component adhesives which are already mixed in their delivered condition. These systems are delivered frozen and start the reaction when they are heated. There are also two-component adhesives which do not have to be mixed beforehand. Here one component is dispensed in the cavity and the other component on the chip module carrier tape. When the chip module is embedded, subsequently the curing reaction starts. This solution is very expensive

Advantage of liquid adhesives:
- Liquid adhesives have been in manifold use in industry for many years. There is a broad pallet of devices and machines for processing, and annually large quantities of these adhesives are issued. The synergy which results from this is undoubtedly a big advantage for liquid adhesives

Disadvantages of liquid adhesives:
- The curing process is connected with the escaping of solvents or different gases and steams, which can unfavourable affect the card materials. Since these gases are usually injurious to health, they have to be sucked off during production which increases the investment costs of the machinery
- Most liquid adhesives have a very thin layer of adhesive between the chip module and card body. This adhesive layer is normally very stiff. This is not a good attribute due to dynamic bending of the smart card. The chip module will pop out of the smart card if the mechanical load becomes to high. With flexible adhesive systems, e.g. heat-activated adhesives, the adhesive layer perhaps would separate on the edges but would still stay in the cavity

4.3   In-mould technology

Injection mould technology is the common name for a smart card finishing technique where the chip module is inserted, held with a mould and afterwards the complete card with the chip module is manufactured with the injection-moulding procedure.

Figure 4.20 shows the structure in principle of the injection-moulding technology.

The mould consists of two pieces and in the upper part is a window of the same size as the contact plate of the chip module. The chip module is a lead-frame chip module (see Section 6.2). The contact plate of the chip module is made from a ferromagnetic material. In the window of the mould a moveable stamp is inserted, which is, for example, wound around with a coil and which can hold lead-frame chip modules an magnetic field. When the two mould parts are separated the chip module is inserted and held by the stamp. Afterwards the mould closes, and using the injecting channel the sealing compound is injected into the mould. After the mould is filled up with the sealing compound and before the sealing compound solidifies, the stamp is presses downward and the chip module is pressed into the sealing compound so that it is positioned at the same level as the card surface. Afterwards the sealing compound solidifies, the mould is opened and the finished card is removed.

The in such a way manufactured injection-moulded card is blank and can be printed on afterwards using the single printing machine.

Beyond the execution described, there are other different solutions possible. One example integrates the printing of the card surface into the injection-moulding procedure. As for the chip module, printed plastic foils are also inserted into the mould or mirror-image printed foils are inserted into the mould, which transfer the image onto the card surface.

Advantages:
- The injection-moulding procedure is a standard industrial procedure, for which broad experience and fully automatic machines are available. From this cost advantages are to be expected
- The raw material for the injection-moulding procedure is plastic granulates. In relation to the laminating procedure the process step of the foil production is not necessary; from this cost advantages are to be expected

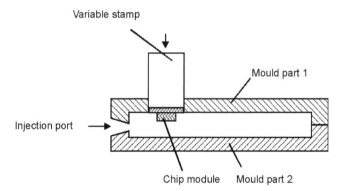

Figure 4.20 Schematic representation of the production of smart cards using the injection-moulding technique

Disadvantages:
- In view of the different coefficients of thermal expansion of the chip module and the card body the card within the chip module area could deform. This causes difficulties afterwards in single printing of the card
- To guarantee the exact position of the chip module in the mould requires special constructional solutions
- The technology is economical for low cost cards like telephone cards. For high-quality cards with many card items (holograms, signature field, embossing and magnetic stripes) injection-moulding technology becomes very complicated

This was an overview of different methods to embed chip modules into a smart card. In the next section the electronic device of the smart card — the chip — is described in more detail.

5 Chips

The integrated circuit is one aspect of the 20th century which has changed almost every part of our daily life. This revolutionary product which is produced from a pot of sand, is at the heart of the smart card and changes a normal plastic card into a smart card. In this chapter we briefly discuss how to produce integrated circuits, with the main focus on integrated circuits for smart cards. Additionally, we over view some basics of integrated circuits technology. For nomenclature purposes 'chip' is used for integrated circuits in this book.

5.1 Basics of integrated circuits technology

Integrated circuits are defined by their specific electrical resistance at room temperature. The typical value is between 10^{-2} and 10^{9} Ωcm, which is between the resistance for metals (10^{-6} Ωcm) and insulators ($>10^{14}$ Ωcm). The electrical conductivity of chips decreases at very low temperature and takes on the characteristics of an isolator. Metals show reverse behaviour. Germanium (Ge) and silicon (Si) are used as material for chips. Only silicon is used for smart card applications.

The characteristics of integrated circuits are strongly tied with the covalent bond of the atomic components.

For covalent bonding elements from the third (III) up to the fifth (V) main group of the periodic table are highly suitable. Silicon stands in the fourth (IV) column of the main group of the periodic table and has four electrons in its outer orbit. To become a chemically stable configuration, the outer orbit of silicon has to be filled up with an additional four electrons. The best partner for the silicon atom is a second silicon atom with its four free electrons. In this case two silicon atoms share two neighbour electrons forming a covalent bonding.

The orbits of the electrons are hybridizing, which means that the bonds are strongly aligned. The consequence is that a silicon atom is surrounded by its neighbour atoms in tetrahedral order as shown in Figure 5.1.

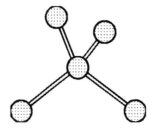

Figure 5.1 Tetrahedral order of the covalent bonds for silicon

With several tetrahedrons deposited together, a regular three-dimensional repeating arrangement develops. This arrangement from unit cells formed in each space direction (*x*, *y*, *z*) with even distances is called a crystal lattice.

The unit cell is a geometric structure which is restricted to flat surfaces. The structure for silicon has the shape of a cube.

There exist three different cubic lattices, as shown in Figure 5.2.

With simple lattices the corner points of the unit cells are occupied by atoms. With the face-centred cubic (f.c.c.) structure additionally the lattice surfaces are filled with atoms. With the body-centred cubic (b.c.c.) structure, similar to the simple lattices, an atom is outside at the corner points and a further atom centred inside the unit cell.

The physical characteristics of crystals are mostly direction-controlled. Therefore the crystal orientations and the planes filled by atoms are indicated by Miller indices (W. H. Miller, 1801 to 1880).

In order to determine the indices, a coordinate system is selected. The axes of the coordinate system run parallel to the edges of the elementary cell. The levels are indicated with three indices *h*, *k* and *l* which are set in round brackets (*h k l*) (see Figure 5.3).

Miller indices are determined by the reciprocal values of the lengths of each axis. The calculation of Miller indices is illustrated by a small example. The example plane is illustrated by the points *A*, *B* and *C* in the coordinate system (see Figure 5.4).

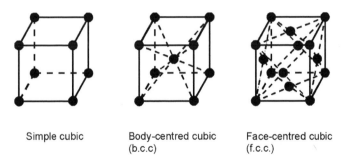

| Simple cubic | Body-centred cubic (b.c.c) | Face-centred cubic (f.c.c.) |

Figure 5.2 Overall view of the three possible cubic unit cells

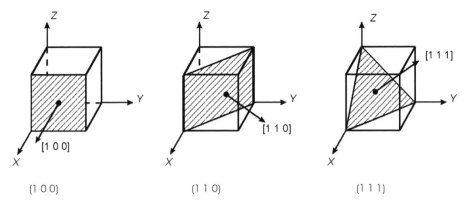

(1 0 0)   (1 1 0)   (1 1 1)

Figure 5.3 Miller indices of important planes in the cubic crystal with the proper crystal orientations

Axis intercepts $x = a/2$, $y = b/3$ and $z = c$; the reciprocal values resulting from this are $1/x = h = 2$, $1/y = k = 3$ and $1/z = 1 = 1$. The Miller indices of this plane are (2 3 1).

A further important value for the classification of crystals is the crystal orientation, which is vertical to the crystal plane. The indexing of the direction is described by three indices u, v and w which are enclosed in square brackets [u, v, w]. For the example illustrated the direction would be [2,3,1]. Figure 5.3 shows the Miller indices for some planes with the appropriate indication of the crystal orientation.

The two most common orientations used for silicon wafers are the [100] and [100] planes, which enable the later separation of the chips from each other on the wafer (also see Section 5.2).

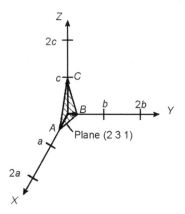

Figure 5.4 The reference system used to define Miller indices

The crystal lattices described up to now are idealized crystals. In reality such crystals do not occur. Real crystals contain lattice defects and foreign atoms in their structure. Some of the possible lattice defects are pointed out briefly.

In Tables 5.1 to 5.3 the three most frequent defect categories — the point defect, the line defect and the volume and surface defect — are specified.

As shown in Table 5.1, the point defects can be divided into different types of defects. With the vacancy or so called 'Schottky defect' imperfection in a crystal is caused by the presence of an unoccupied lattice position. If additional atoms are present in an otherwise complete lattice ('Anti- Schottky defect'), this is called an interstitial atom defect. Frenkel-defect pairs are a combination of vacancy and interstitial atoms. If impurity atoms are present in an otherwise complete lattice, two different defects are possible. First, a so-called interstitial impurity which describes an additional atom, which does not belong to the host crystal, between positions in the occupied lattice is possible. In addition, if the impurity atom replaces or substitutes an atom present in the lattice this is termed a substitutional impurity.

Table 5.1 Overall view of point defects

Point defects				
Vacancy atom	Interstitial atom	Vacancy and interstitial atom	Impurity atoms	
Schottky defect	Anti-Schottky defect	Frenkel-defect pair	Substitution impurity	Interstitial impurity

Apart from the point defects there are also so-called line defects, which are also divided into two different types of defects.

If the atomic planes end as wedges in the crystal, then this defect is called an edge dislocation. The sliding direction runs perpendicular to the dislocation line. With the screw dislocation the crystal lattice is displaced parallel to the dislocation line by one plane level. The distance of the displacement is called the Burgers vector (b) (see Table 5.2).

Table 5.2 Outline of the lattice defects as line errors

Line dislocations	
Edge dislocation	Screw dislocation

Table 5.3 shows surface and volume defects. Among surface defects the stacking faults rank as foremost. Here, additional planes are inserted or removed from the lattice. With the second defect, grain boundaries between differently oriented crystal areas occur. In the case of volume defects this mostly concerns pipe (pores) and/or segregation of a second phase in the solid state. These can develop, if oxygen is taken up from the melt during production of the single-crystal silicon.

All the mentioned lattice defects can negatively influence the mechanical and electrical characteristics of the chip.

Table 5.3 Surface and volume lattice defects

Surface and volume defects	
Stacking fault	Grain defects
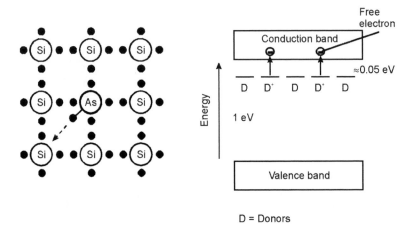	

However, one of these defects is caused deliberately during crystal production, i.e. the direct introduction of specific atoms into the lattice structure. This process is called doping. The procedure is used to produce electrical conductivity in semiconductor material. Depending upon the type of foreign atoms used we obtain p-type (doping with acceptors) and n-type (doping with donors) semiconductors.

Donors elements used include phosphorus (P), arsenic (As) or antimony (Sb). Boron (B), aluminium (Al), gallium (Ga) and indium (In) are used as acceptors.

The electrical behaviour of semiconductors is most easily illustrated with the so-called energy band model:

The energy band model proceeds from an ideal single crystal, in which positively charged atoms sit in a periodic crystal lattice. Due to the electrostatic interactions of electrons with neighbouring atoms the electron levels split. The two energy levels are separated by a restricted area, a so-called energy gap or band gap.

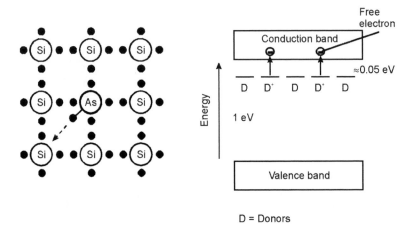

D = Donors

Figure 5.5 Energy band scheme for n-type semiconductors

The two energy levels are separated into a valence band and a conduction band. The conduction band is only partly filled. The fully filled valence band is separate from the conduction band by a band gap which is for example 1 eV.

If the semiconductor was doped with donor atoms then an n-type semiconductor develops, as shown in Figure 5.5. The principle being that a free electron can move from one atom to the next.

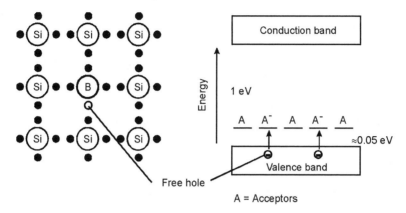

Figure 5.6 Energy band scheme for p-type semiconductors

However, if the material was doped with acceptor atoms then a p-type semiconductor develops as shown in Figure 5.6. Here, the principle is that the missing electron (hole) moves through the lattice.

In a functional semiconductor (chip) the n- and p-zones have to be united on the material. Using the basis of a bipolar transistor, we briefly demonstrate, how the n- and p-zones are distributed.

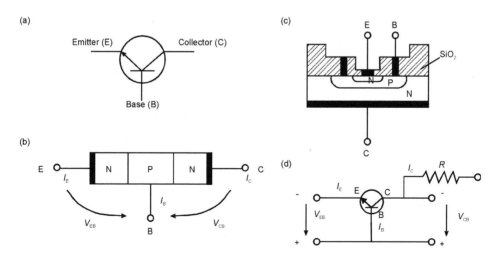

Figure 5.7 Different models of an n—p—n transistor
(a)   Circuit symbol of an n—p—n transistor
(b)   Schematic arrangement of the n—p—n layers
(c)   Structure in principle of an n—p—n transistor
(d)   Basic circuit of an n—p—n transistor

Figure 5.7 shows the principles of a bipolar transistor and a sample circuit. In this circuit the voltage V_{EB} is set on the emitter/base junction in the forward direction. The blocking voltage V_{CB} occurs between the base and collector of the transistor. Hence, electrons flow from the emitter to the base. Here the current I divides itself into a minority base current I_B and a majority collector current I_C. If the base zone is enough the collector current has approximately the same value as the emitter current. The current amplification factor is approximately one. Since practically the same current is at the input with a low input impedance and a small voltage V_{EB}, at the output, however, because of the high output impedance effects a high voltage V_{CB} causes a voltage amplification.

A large range of different circuits needs to be described, but this should be enough to give a short overview and introduction to the basic mechanisms of the semiconductors.

Following these basics, the production processes for semiconductors and chips are described in the following Section.

5.2 Production processes for chips

In this Section we describe the production steps of a wafer in order to obtain an overview of the complexity of the procedure for the chip, which is at the heart of each smart card, to be ready for application in the chip module and afterwards in the smart card.

Simply described, a wafer is a round disc of silicon on which through several process steps the electrical circuits of the chips are implemented.

Figure 5.8 shows an overview of how the individual process steps for a chip develop from quartz sand.

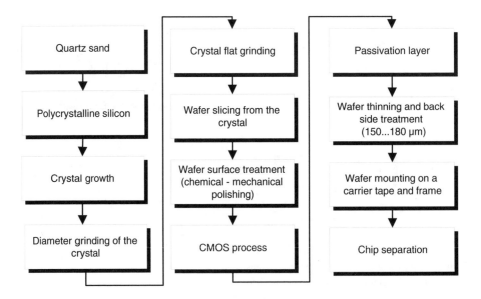

Figure 5.8 Flow scheme for the individual production steps from quartz sand to the chip

All these process steps are executed in a so-called clean room. A clean room is an air-conditioned fabrication area where the air temperature and humidity are held at a constant value. A preset concentration and size of contamination (e.g. dust particles, human hairs or smoke particles) may not be exceeded in ambient air. The air quality is divided into different number classes. The air in a normal city with all its smog and dust can have more than five million particles per cubic foot which are larger than 0.5 µm in diameter. This concentration would correspond to a class number of five million. The relationship between particulate diameter and density is defined by Federal Standard 209E.

For a wafer fabrication process in a clean room, a class number of 0.1 to 10 with a maximum particle diameter of 0.3 µm is required. Figure 5.9 shows a typical clean room for wafer production.

Figure 5.9 View of a clean room for wafer production (source: Infineon)

The first manufacturing step from Figure 5.8 is production of silicon from quartz sand. For this normal quartz sand (SiO_2) is reduced according to Equation (5.1) to produce raw silicon.

$$SiO_2 + 2C \Rightarrow Si + 2CO \uparrow \tag{5.1}$$

Through a chemical process the raw silicon is converted to polycrystalline high-purity silicon. The degree of purity is 99.999999 %. In the next step the polycrystalline high-purity silicon is doped and changed into a single-crystal structure through melting and solidifying. The most common procedure for the production of mono-crystalline silicon is the Czochralski crystal-growing system as shown schematically in Figure 5.10.

The polycrystalline high-purity silicon is brought into a rotary chuck, which consists of quartz. It is melted with radio-frequency (RF) waves, (melting temperature for Si: 1412 °C). Next, a seed crystal, which is fastened on a rotary chuck, is positioned to just touch the surface of the melt. The seed crystal is a single crystal which functions as centre of crystallization. The seed crystal influences the crystal orientation as well as other failures, which can develop in the new single crystal.

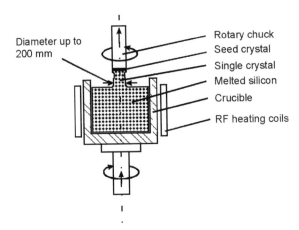

Figure 5.10 Czochralski crystal-growing system

At the beginning of the growth of the single crystal the seed crystal is quickly pulled out of the melt, as a result the single crystal obtains the form of a bottle-neck. This bottle-neck shape is good for the single crystal as the initial lattice dislocations run out quickly to the margin and end there. If the dislocation-free growth of the single crystal begins, the pullout rate of the seed crystal is reduced. At this stage the diameter of the single crystal becomes larger, until the desired diameter is achieved.

To always keep the growth zone in the same area, the crucible is adjusted upward at the same rate as the melt is used up. The entire process is undertaken in a low-pressure atmosphere using argon gas.

A disadvantage of the Czochralski process is the high concentration of oxygen in the single crystal. This oxygen results from the contact of the melted silicon with the wall of the quartz crucible (see Equation (5.2)).

$$SiO_2 + Si \Leftrightarrow 2SiO \uparrow \tag{5.2}$$

A majority of the SiO evaporates at the surface of the melt; however, a small part is built-in during the crystallization.

The oxygen atoms have a negative effect on the electrical conductivity of the silicon crystal. If the single crystal is heated up to 650 °C and afterwards cooled down spontaneously, then the oxygen pollution can be reduced.

The oxygen atoms also have an advantage, as unwanted lattice defects and metal pollution preferably deposit themselves at the oxygen in the single crystal. This advantage is used for collecting of pollution in the silicon wafer and is termed 'gettering'.

The pulled single crystals can achieve diameters up to 200 mm (8 in.) and lengths up to 2 m (see Figure 5.11).

Figure 5.11 From the top right: polycrystalline parts, single crystal, polished wafer in different sizes, wafer in transport containers (source: Infineon)

Next, the very long single crystals are sawed into pieces of approximately 50 cm length. The parts are examined according to the agreed specifications, for example the specific resistance; if the parts fail they are requested to be sorted out.

After pulling the single crystal it may possibly not be exactly round or have the same diameter over the whole length. Therefore it is ground with a diamond grinder to achieve the required diameter.

As already mentioned in Section 5.1, there are different crystal planes in the single crystal. The crystal orientation can be determined by the so-called Laue procedure using sharp bundled X-rays. They are directed towards the single crystal. With the help of the developing interference pattern the orientation of the crystal can be determined.

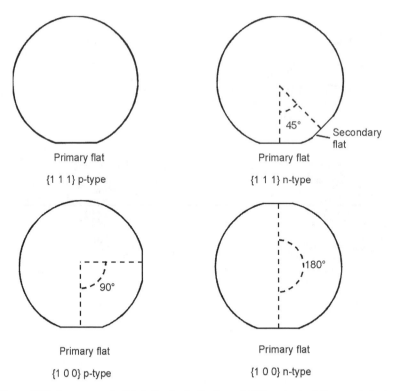

Figure 5.12 Arrangement of flats for the indication of different orientations and doping

For use in later manufacturing steps, the single crystals are provided with a marking, the so-called 'flats'. A differentiation takes place between the 'primary flat' and 'secondary flat'. Figure 5.12 shows various combination options (whereby the doping and orientation of the wafer are characterized).

Primary flats are used for the mechanical adjustment of the wafer and for the adjustment of the chips, as these must be aligned relative to the crystal orientation.

For further production steps the wafers are sliced. The 50 cm-long crystal is divided into several 0.5 mm thick single wafers. For this process special inside-diameter saws are used. The outside edge of the saw blade is tightened onto a circular frame. Extremely thin saw blades up to 120 µm can be used, which reduce the kerf (cutting width) size.

Figure 5.13 schematically shows an inside-diameter saw. The saw blade is coated with diamonds.

For the later use of the wafer in MOS (metal-oxide semiconductor) technology the sawing direction of the {100} material is in the high symmetry plane, while {111} crystals are sawn with a small angle deviation in relation to the {111} plane.

Individual wafers are available. Before their suitability for chip production, the next process step is to give the wafer a surface treatment and free it of irregularities and sawing damage.

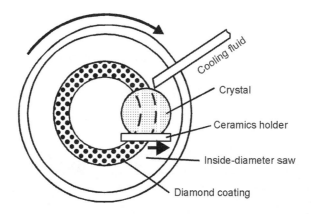

Figure 5.13 Principle of an inside-diameter saw for sawing a crystal

During sawing the wafers were mechanically damaged and contaminated. In the upper zones near the surface there are lattice defects, which are termed 'hair-cracks' or 'micro-cracks'. These cracks disturb the further wafer processing. In order to eliminate such hair-cracks the wafer is cleaned first with high-purity water and polished afterwards. During polishing the wafer is pressed on a slowly rotating holder and moved back and forth with an eccentric disc. Using several polishing steps the grain size of the polishing material (Carborundum in water suspension) is reduced gradually. The remaining thickness of the wafer is around 360 µm.

Afterwards the sharp edges of the wafer have to be grounded in order that the wafer has rounded edges. This is necessary as during the later wafer processing chipping parts can damage or contaminate other wafers. Additionally cracks from the edge continuing into the wafer have to be avoided.

In order to eliminate this source of lattice damage the wafer is etched with alkaline (potassium hydroxide) or acid media (nitric acid and hydrofluoric acid).

After a number of final inspections the wafers are ready for further wafer processing.

A range of process steps follow, which change the silicon wafer into a working chip with a multiplicity of transistors and junction lines. In the following only a few necessary process steps of CMOS (complementary metal-oxide semiconductor) technology are listed:

- Photoresist coating of wafers using spinners
- Soft bake wafers in oven
- Align wafer and mask exposure to light source. (A mask is a glass plate covered with an array of patterns used in the photo-masking process. Each pattern consists of opaque and clear areas that, respectively, prevent or allow light to pass. The mask surface may be emulsion, chrome, iron oxide, silicon, or a number or other materials)
- Develop and remove unexposed photoresist
- Remove exposed oxide with wet etch or dry etch process
- Apply dopant in diffusion furnace or ion implanter
- Add deposition layers in CVD (chemical vapour deposition) furnace or epitaxial reactor (above steps repeated as required)
- Metallization to connect circuit layers

- Passivation to form protective coating

In the last process step the passivation layer is applied on the chip surface. Subsequently, the chip pads must be free-etched again, so that the chip can be contacted electrically later.

Figure 5.14 shows a CMOS inverter after the last process step of the passivation process.

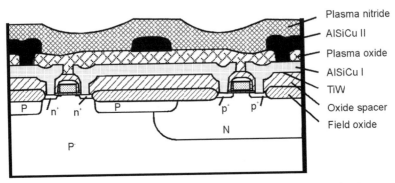

Figure 5.14 Structure of a CMOS inverter after applying the passivation layer

The passivation layer represents an important interface for the chip to the later chip module. The function of the passivation layer is to protect the circuits underneath against humidity, corrosive gases and mechanical loads.

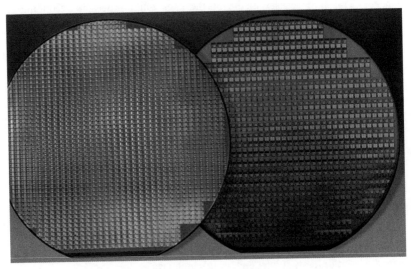

Figure 5.15 Two finish produced wafers (source: Infineon)

Many passivation layers consist of plasma nitride which is brittle and inclined to cracking. Plasma oxide can also be used as the passivation layer, it has a stronger water permeability than the plasma nitride. The advantages of both materials can be combined by separating plasma nitride on top of a layer of plasma oxide. Figure 5.15 shows two finished wafers.

Now the wafer can be used for mounting in a DIP (dual in-line package) —it only needs the final electrical outgoing test.

The electrical outgoing test normally follows the testing procedure shown in Figure 5.16.

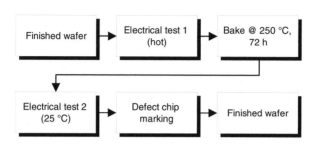

Figure 5.16 Sample for the electrical wafer outgoing test

During electrical wafer testing the performance and circuit functioning is tested. The finished wafer is brought to a wafer tester where each chip on this wafer can be contacted electrically with a computer via the pads on the chip. For the first test step the electrical testing is undertaken at a high temperature. Next the wafer is put into a heated oven (250 °C) for 72 h. This step is the so-called wafer bake. The bake procedure is performed to determine the data retention capabilities of the chip. Afterwards the wafer is tested at room temperature to see how the bake influences the quality of the complete wafer. The computer used to perform the electrical test stores the results of the test. In the next step the defect chips on the wafer are marked with an ink dot (see Figure 5.17). Therefore during the package of the single chip, the defect chips are noticed and can be avoided. Also sometimes the position of the defect chip is electronically stored on a floppy disk — a so-called wafer map. This information can also be used later for the chip package machine to detect the defect chips.

For a smart card module the wafer has to be thinned from its usual thickness of 360 µm down to approximately 180 µm. Only with thinned chips can the total thickness of the chip modules be kept low (also see Section 6). After the thinning process the wafer is fixed on an adhesive tape of approximately 70 µm thickness. The tape is stuck to a frame, in order to simplify the later handling of the wafer (see Figure 5.17).

Now the chips on the wafer must be separated along the sawing streets (scribe lines) through mechanical processing (see Figure 5.18).

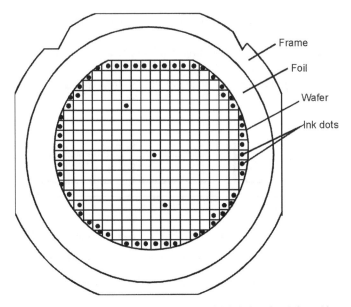

Frame
Foil
Wafer
Ink dots

Figure 5.17 Sawn wafer on a sticky tape with ink dots for defect chips

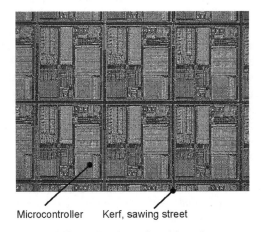

Microcontroller   Kerf, sawing street

Figure 5.18 Details of a wafer with sawing streets

Saws are used for the separation. The procedure is called 'dicing'. The saw blade consists of a 25 µm thick steel plate coated with diamonds (see Figure 5.19). The saw blade receives its stiffness from the high rotation speed. During the dicing the saw blade is cooled with deionized water. The saw cuts through the wafer until it scratches a few microns into the tape. The dicing has to be done very careful as it could cause back side chipping which may influence the later mechanical quality of the chip (see Figure 5.20).

Figure 5.19 Sawing plate of a wafer with water cooling (source: DISCO)

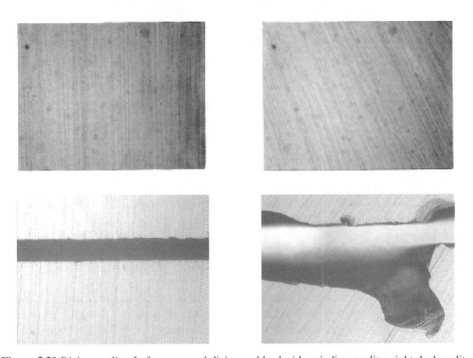

Figure 5.20 Dicing quality. Left: very good dicing and back side grinding quality; right: bad quality for back grinding with scratches and chipping in the area of the dicing (bottom right) (source: DISCO)

After these various production steps and the appropriate electrical and optical checks the chip is prepared for application into the chip module.

5.3 Paper-thin wafer

As described in the previous chapter, the used wafer thicknesses for chip modules and also for contactless cards are between 150 and 185 µm. With this thickness the chip is stiff and very inflexible. But if the wafer is thinned down to a thickness under 50 µm the wafer itself and especially the chip becomes flexible as can be seen in Figure 5.21.

Figure 5.21 Flexible wafer (source: DISCO)

Thin chips between 50 and 20 µm, are also called 'paper-thin wafers' and can be used to make very thin chip modules (<400 µm) and also very thin and flexible contactless smart cards. Here smart cards with a thickness of 250 µm could be made.

At the moment there are a number of different ways being developed to produce a paper-thin wafer. The DISCO company together with Toshiba and Lintec developed a thinning process called DBG (Dicing Before Grinding). In the following DBG is described in more details (see Table 5.4).

In the first process step a cut approximate 50 µm deep is made into the thick wafer. After dicing a carrier tape is mounted to the active side of the chip (pad side) and turned around in order that the back side of the wafer shows to the grinding tool. Then the wafer is back-grounded down to a thickness of 50 µm. After the grinding process the individual chips on the wafer are separated. Now only the wafer has to be flipped again, so that the active side (pad side) of the chip is upside. For this a carrier tape is mounted on the back side of the chips, now that the chip is between two tapes like a sandwich. The system is turned around (flipped) and the tape on the top side of the wafer is removed. The chip is now in the right position for the later assembly process and is put into a transport box.

Table 5.4 Flow for DBG (Dicing Before Grinding)

DBG process steps		
#1	Half-cut dicing	
#2	Tape lamination	
#3	Back grinding & in-line transfer unit	
#4	Mounter/remover (with flip function)	
#5	Wafer in transport box	

Figure 5.22 shows a fully automated DBG in-line system with back grinder, transfer unit and tape mounter/remover.

Figure 5.22 DBG in-line system (source: DISCO)

Nowadays the process of making thin wafer is very well known. But the further handling of paper-thin chips like die placing has not been completely worked out. This will take another year until machines are available for mass production.

5.4 Types of memory chips

Section 5.2 dealt with the fundamentals of chip production; different memory types (see Figure 5.23) and their interplay are briefly described in this section.

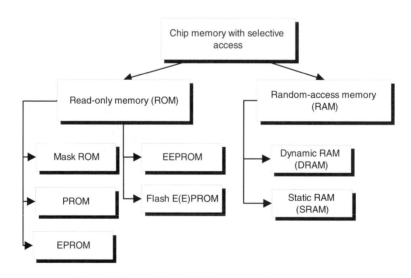

Figure 5.23 Overview of different types of chip memories

Selective memory access allows the possibility to have access to every memory cell. If the chip memory is compared with a chessboard, then each field of the chessboard corresponds to a memory cell. On each field information can be stored (1 or 0).

Chip memories are divided into two memory types: only readable memory, so-called ROM (read-only memory), and write/readable memory, so-called RAM (random access memory).

Readable memories (ROM) are predominantly appropriate only for read access. ROM memories are non-volatile memories, this means that the memory does not lose its stored information when it loses its power as is the case, when the smart card is removed from the terminal.

The mask ROM memory is the memory of the smart card chip where the operating system of the smart card is stored. During the chip production the operating system is brought into the chip memory array section via a mask. This process is irreversible and a fixed part of the memory circuit.

The PROM is a programmable ROM and can be programmed uniquely and irreversibly by the user. After the programming is finished the PROM changes into a ROM.

The EPROM (erasable programmable read-only memory) is electrically programmed by the user like a PROM. The whole memory can be deleted by UV irradiation. However, this is not desired for a smart card application, therefore EEPROM (electrical erasable PROM) is used. EEPROM can be electrically programmed like an EPROM and also deleted electrically. The time, in order to write information into the memory cells, lasts longer with an EEPROM than with a RAM.

Also the flash E(E)PROM is electrically programmable like an EEPROM, but the memory content can be deleted electrically very fast (flash), which is not possible with the EEPROM. Flash E(E)PROM is normally used in smart card applications to test the operation software or for field tests.

Turning to the second group of chip memory with selective access — write/readable memory RAM (random-access memory): With RAM, contrary to EEPROM, a very fast writing and reading of information from each memory cell is possible. RAM is a volatile memory and loses its data contents after the power to the smart card goes off.

The difference between DRAM (dynamic RAM) and SRAM (static RAM) exists in the memory principle.

With DRAMs the information is stored in the chip in condensers. Since with all condensers leakage currents occur, the information would be lost after some time, therefore the information must be refreshed periodically — about every 2 ms.

Static RAM memories (SRAMs) store their information in so-called 'flip flop' trigger circuits. Since a 'switch' is changed between status '0' off and '1' on the memory content does not have to be refreshed as with DRAM. The disadvantage of SRAM is the increased power demand which is needed to activate the 'switch'.

Some of the mentioned chip memories (ROM, EEPROM and RAM) are used in a smart card chip.

There is a difference between a pure memory chip smart card and a microprocessor smart card. The simplest smart card, related to functionality and cost, is the pure memory chip smart card.

Figure 5.24 shows the most substantial constituents of a memory chip as a block diagram.

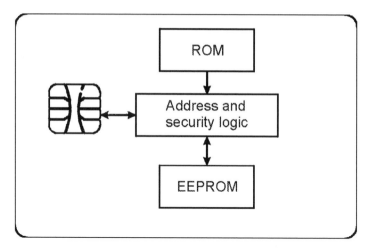

Figure 5.24 Block diagram of a pure memory smart card

The necessary data for application are stored in the EEPROM. The data of the chip module are transferred to the terminal over a safety logic between the EEPROM and the input/output (I/O port). The identification data for the safety logic is stored mostly in the ROM. Figure 5.25 shows memory chip SLE4436E from the Infineon company.

Memory chip smart cards are used because of their low production costs for calling cards or health cards.

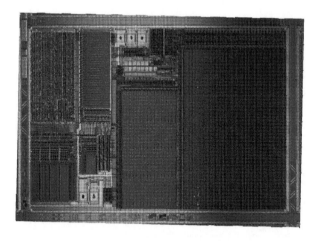

Figure 5.25 Memory chip SLE4436E (source: Infineon)

If the application, e.g. for mobile phones, conditional access or as an electronic purse, needs higher flexibility and a higher safety level then smart cards with inserted microprocessors are used. Figure 5.26 shows the simplified block diagram of a microprocessor chip.

The focal point is the CPU (central processing unit). The CPU takes over the central flow control and coordination of information in a microprocessor chip. It is supported by the RAM, ROM and EEPROM. The instruction sets of the Motorola core 6805 and Intel 8051 Architecture were established for the CPU.

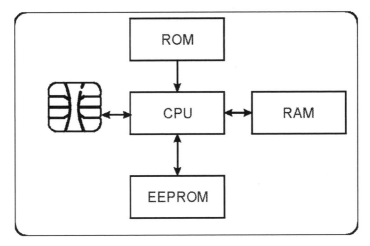

Figure 5.26 Block diagram of a microprocessor smart card

The operating system of the microprocessor is stored in the ROM. The EEPROM is used for the long-term storing of data. During operation of the microprocessor smart card the RAM functions as a working memory as when the power supply is switched off the RAM loses the stored data.

With smart cards, where personal data are needed, e.g. name, address or customer number, the information is written into the EEPROM during personalization of the microprocessor chip.

Communication between the chip (memory or microprocessor) and smart card reader (terminal) is undertaken by an input and output interface which are provided on the chip module. These are described in the next chapter.

6 Chip modules

In the preceding chapter the function and manufacturing process of the chips were described. In this chapter we demonstrate how the chip can be protected against external influences and how it is electrically connected to the external world.

6.1 Why chip modules?

Approximately 20 years ago the first smart cards were brought on the market. Since this time chip modules — the most substantial component of the smart card — have been continuously further developed.

How did chip modules establish themselves in such a way, and what are they good for?

If one regards the production process of chips (also see Section 5.2) — enormous difference to the production process of a smart card becomes recognizable. With chips the highest demands are made for cleanliness of the production plants. With smart card production the cleanliness of the production plant is likewise important, but such a high level of cleanliness is not necessary.

Also the process materials used such as fluorine and other corrosive chemicals, which are used during chip production, do not arise in smart card production. Supply of a pure environment and processing chemicals with their pertinent disposal is very cost-intensive. Therefore it is not advantageous to merge chip production and chip module production.

Another aspect for the application of a chip module is found in the fact that the chip has only very small electrical connection areas, so-called pads, with a size of $100\ \mu m * 100\ \mu m$. In order to create a reasonable electrical connection to the smart card reading devices an increased contacting surface has to be created in order to enable data exchange between readers and chips. Here the chip module with its large contact areas is ideal.

Besides increased size and resistance of the contacting surface, the chip module serves as mechanical protection for the chip. The chip is enclosed at the rear side of the contact surfaces. Hence the bonds, which make the electrical connection between the chip pads and the external contacts of the chip module, are protected. The gold wire used for the connection ($25\ \mu m$ in diameter) is thinner than human hair and very sensitive to mechanical loads.

Also the chip module reliably protects the chip against chemical influences such as corrosion.

A final advantage is that the chip module offers the possibility, independently from the different manufacturing methods of chip production, to develop new products and product properties in chip module and smart card production. The interfaces between the chip and chip module, for example chip passivation, position of the pads and the substrate of the chip module itself, must be coordinated to one another.

The following sections show an overview of the different chip modules and their structure and manufacturing process.

6.2   Different types of chip modules

In the past few years a multiplicity of different chip modules for smart cards have been established. Chip modules are frequently defined as COB (chip on board) or COT (chip on tape). Chip modules can be divided into different groups.

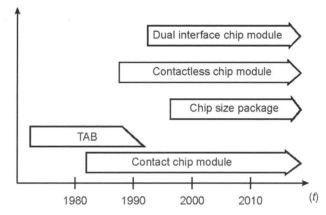

Figure 6.1 Overview of the temporal development of different chip modules

In Figure 6.1 represents a temporal overview of development of the different chip modules.

One group of chip modules termed TAB (tape automated bonding) modules were used with the first smart cards for telecommunication applications.

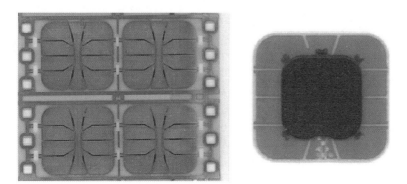

Figure 6.2 Left: four microprocessor chip modules each with eight contact areas in the carrier tape (front). Right: the punched-out chip module with an injection-moulded encapsulation (rear side)

A further group in use presently are chip modules with six and eight contact areas. One can distinguish the technology used for encapsulation of the chips. Figure 6.2 shows a chip module where the encapsulation was covered with an injection-moulded glob top. Companies such as Atmel and Hitachi offer this type of chip modules as package for their chips.

A further option for making the encapsulation involves covering thermally hardening epoxy resin with ultraviolet (UV) rays. Chip modules of this design are predominantly offered by NedCard, Philips and ST Microelectronics with chip modules such as D40, D45 and D30 and by Infineon with chip modules such as M3.2, M2.2, M4.3, M5.1 and M5.2 (see Figure 6.3).

Also, the contact chip modules differ in the number of contact areas; they possess either six or eight contact areas (see Figure 6.4).

Figure 6.3 Memory chip module with UV-hardening epoxy resin (M2.2) and microcontroller chip module with stiffening frame (M4.3) (source: Infineon)

 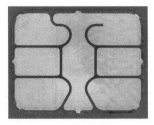

Figure 6.4 Contact sides of two chip modules. On the left a chip module with eight contact areas. On the right a chip module with six contact areas (source: Infineon)

The selection of the chip module (six or eight contact areas) depends only on the size of the chip used. Chips with a size up to approximately 5 mm² can be built into a chip module with six contact areas. Chips with larger dimensions are used in a chip module with eight contacts. The maximum size of chips is approximately 32 mm².

For the user of the smart card the visible parts of a chip module are the golden contact areas. These contact areas represent the smart card interface with the external world. The chip is connected to the smart card reader via the bond wires and the contact areas. To ensure the contact points are at the same position in every smart card, an international standard, ISO 7816 applies. This standard describes the minimum contact area and its position on the smart card (also see Figure 1.3).

The functions of the contacts C1 to C8 are also defined in ISO 7816 (see Figure 6.5).

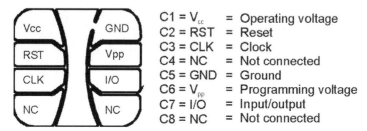

$C1 = V_{cc}$ = Operating voltage
$C2 = RST$ = Reset
$C3 = CLK$ = Clock
$C4 = NC$ = Not connected
$C5 = GND$ = Ground
$C6 = V_{pp}$ = Programming voltage
$C7 = I/O$ = Input/output
$C8 = NC$ = Not connected

Figure 6.5 ISO contact areas for a chip module with eight contacts

Contacts C4 and C8 are not connected to the chip in regular use but are reserved for future or special applications. In today's chips, the programming voltage is produced chip-internally hence C6 is also not occupied.

For the contactless card there are — as with the contact chip modules — two different module types: chip modules with chips which are encapsulated with epoxy resin, and chip modules which protect the chip with an encapsulation manufactured using the injection-moulding technique. Figure 6.6 shows the MCC1 and the MCC2 chip modules for the contactless smart card from the Infineon company. Philips offers a chip module named MOA2 which is structured according to the same principle.

Figure 6.6 Contactless chip module with injection-moulded encapsulation and lead-frame contacts (source: Infineon)

Contrary to the contact chip modules the contactless chip modules only need two electrical contacts. These are needed for the later electrical link of the antenna to the chip module. The contact areas, however, are not determined by the contact chip modules hence there is no standard for size and position.

To increase functionality, the advantages of contact chip modules have been combined for some time with the advantages of contactless chip modules. Hence a new chip module called dual-interface chip module was created. Figure 6.7 shows an example of a dual-interface chip module from the Infineon company.

Figure 6.7 Dual-interface chip module seen from the contact side (left) and the rear side (right) with the two antenna contacts (source: Infineon)

Here the eight ISO contact areas are situated on the front of the chip module. On the rear side, arranged to the card body side, are the two contacts for the antenna present in the card. The encapsulation can take place again with the injection-moulding technique or with epoxy resins.

Thus the chip in a dual-interface chip module is able to exchange data with a reader via the ISO contacts or contactless over the antenna. Figure 6.8 shows the structure of a dual-interface chip module. Infineon, ST Microelectronics and Philips offer dual-interface chip modules.

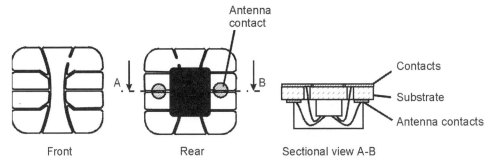

Figure 6.8 Schematic structure of a dual-interface chip module

Finally, the so-called transponder chip module is shown here (see Figure 6.9). The transponder chip module is an alternative to the conventional contactless smart card as the antenna for data communication is located on the chip module and not in the card body.

The antenna is used for contactless transfer of data between the smart card and reader (see Figure 6.10).

Figure 6.9 Transponder chip module with antenna (front, on the left) and encapsulation (rear side, on the right) for contactless data communication

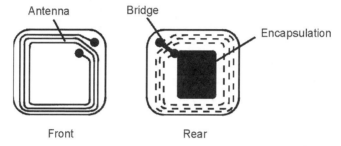

Figure 6.10 Schematic structure of a transponder chip module with integrated antenna on the chip module front

The transponder chip module can be additionally equipped with ISO contacts for contact applications. The transponder chip module can be built into the card body with the help of the mounting technique (see Section 7). The working range between contactless smart cards with a large antenna embedded in the card body and the reader device is about 10 cm. The usage of the transponder chip module with its smaller antenna size causes a reduction of the working range up to 3 cm due to lower transfer of energy between smart card and reader.

6.3 Structure and materials for chip modules

Section 6.2 presented a short overview of the different most commonly used chip modules. In this section the structure and the materials used are described in more detail.

Figure 6.11 gives an overview of the different components of a chip module, which are described in more detail in the following.

In Germany the first calling cards were equipped with TAB modules (tape automated bonding). Figure 6.12 schematically shows the structure of a TAB module.

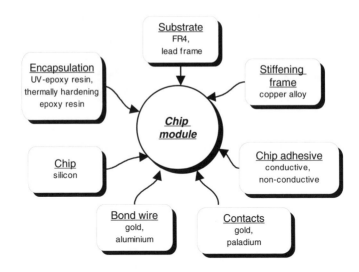

Figure 6.11 A selection of different materials for a chip module

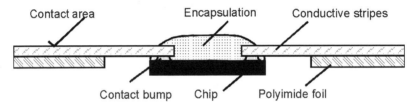

Figure 6.12 Schematic structure of a TAB chip module

Initially during the TAB procedure metallic peaks (bumps) are applied galvanically on the chip pads. Afterwards the conductive stripes (wire frames) and the chip pads are connected in one processing step by soldering with a hot stamp. The link conductive stripes are bound on a polyimide foil (trade name: Cape tone) connected to the ISO contact areas. Finally the chip and the junction points are covered with a sealing compound. The TAB connection technique is mechanically very stable, in addition the thickness of the chip module is lower than the chip modules currently in use.

In spite of these advantages TAB modules are no longer used due to cost reasons.

Instead the chip modules which are schematically shown in Figure 6.13 and Figure 6.14 are used today.

Copper is used as base material for the contact plate. A thickness of 35 µm or 70 µm is used which is improved galvanically by a nickel/gold coating. The layer thickness of nickel is between approximately 2 and 4 µm and that of the gold layer approximately 0.1 µm.

Galvanically applied nickel layers are mechanically very hard. Gold coating is used as gold is chemically very steady and possesses very good electrical conductivity.

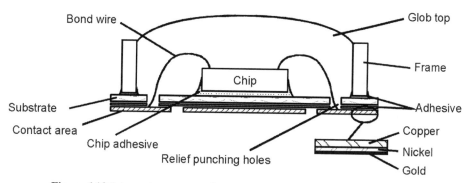

Figure 6.13 Schematic structure of a contact chip module with stiffening frame

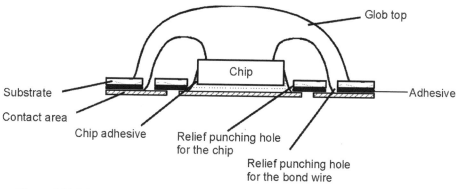

Figure 6.14 Schematic structure of a contact chip module with a drop-shaped encapsulation

From cost reasons the gold layer is kept as thin as possible. The layer thickness, however, should be at least thick enough that no pores exist in the gold layer. With such damage present, humidity can accumulate between the gold and nickel layer. The result is that the two metals (gold and nickel) form a galvanic cell, whereby gold forms the cathode and nickel the anode. Oxides and sulphides form under the gold layer and lead to deterioration of the electrical characteristics of the contacts.

The nickel barrier layer between the copper and the gold is also necessary in order that copper does not diffuse into the gold layer and vice versa. Otherwise the contacts are more easily inclined to corrosion formation when humidity occurs.

For the later wire bonding (see Section 6.4) the copper layer is improved with a nickel/gold layer in the bond area on the back side of the ISO contacts. Here the nickel layer again serves as a barrier in order that copper does not diffuse into gold. If copper arrives at the bond contacts, the adhesive strength of the gold wire is strongly lowered and the reliability of this connection reduces.

The layer strength of the gold/nickel layer also depends on the required abrasion firmness. A high abrasion firmness is necessary in order that the gold/nickel layer is not worn out by the contacts of the reading head when frequently put into a smart card reader. If bright copper shows up, corrosion of the contact areas is enabled and electrical functionality is influenced. Alternative to the gold layer recently palladium has been used for surface refinement.

Glass fibre-reinforced polymers, so-called FR4 material are used as carrier substrate for the coated copper foil. These have a thickness between 80 and 120 μm. For cost reduction in some cases punched metal carriers (so-called lead frames) are used. FR4 material can be obtained from the Risho company, where as punched metal carriers are sold, for example, by the Heraeus company.

The substrate is provided with free punching holes, in order to connect the chip pads with the contact plate through the bond wires.

With some chip modules there is free punching for the chip within the chip area (see Figure 6.14). Thus the chip is glued directly on the improved copper plate instead of on the substrate (see Figure 6.13). This is a common method in order to reduce the total module thickness. Manufacturers for chip module tapes include MCTS, IBIDEN, Cicorel and Multitape.

Different bond wires are in use depending upon the used bond process, e.g. gold wires with a diameter from 24 to 35 μm (ball—wedge) or aluminium wires in the so-called wedge—wedge bonding process.

With chips of a size larger than 20 mm² sometimes it will also be necessary to apply a stiffening frame additionally on the substrate. Thus increases the total rigidity of the chip module, in order to protect the chip against breaks in everyday use. A copper alloy is usually chosen as material for the stiffening frame.

For packaging of the chip and its bond wires encapsulation resins on epoxy base are used. These are hardened by temperature influence or UV light. For sealing compounds there are a multiplicity of suppliers available, e.g. Dexter, DELO or Panacol.

Certain criterion have to be considered in selection of the sealing compounds. Among these is the shelf-life of the material. The working time (dropping time) should be as long as possible, otherwise frequent change of the sealing compound material is necessary and the consequence is unnecessary production downtimes. The moisture absorption and ion concentration of the material are likewise very important. If, for example, many free ions exist in the sealing compound this may cause increasing corrosion formation. High humidity concentration may lead to the so-called popcorn effect during the later embedding into the card body, this occurs under temperature influence.

The popcorn effect means a spontaneous lift-off of the sealing compound from the chip surface or from the chip edges. The separation is caused by the spontaneous evaporation of the enriched humidity between the sealing compound and chip surface. This is caused by the heat energy linked into the chip module over the welding stamps. During the spontaneous separation of the sealing compound the bond wires often split. In order to avoid the popcorn effect, low moisture absorption of the sealing compound is important.

Instead of dispensing liquid resins the injection-moulding technique also may be used — it is applied with a commercial DIP (dual-inline package) for the encapsulation of the chips. The chip and the bond wires are enclosed with a thermosetting polymer in the injection-moulding procedure.

For purely contactless chip modules normally a punched lead frame is used as substrate and the moulding technique is used as encapsulation.

For dual-interface chip modules an additional second contact plate is applied on the substrate contrary to contact chip modules. Thus the antenna contacts can be led from the chip to the external antenna when arranged on the side of the card body (see Figure 6.15).

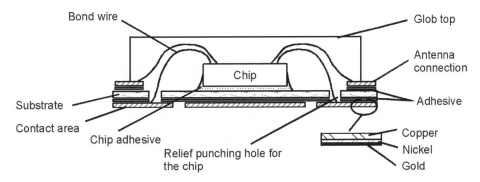

Figure 6.15 Schematic structure of a dual-interface chip module

As a geometrical dimension for chip modules with eight contact areas an exterior punching measure of $X_{M1} = 11.8$ mm $*$ $Y_{M1} = 13.0$ mm (see Figure 6.16) is established. For chip modules with six contacts an exterior punching measure $X_{M1} = 8.32$ mm $*$ $Y_{M1} = 11.0$ mm is very common.

For the carrier tape substrate and contact area thickness a d_T of 160 µm for contact chip modules and at least 200 µm for dual-interface chip modules is established. The carrier tape thickness d_T for contactless chip modules is about 110 µm.

The total thickness d_G varies depending on the structure of the chip module and lies between 550 and 620 µm for contact chip modules. With contactless chip modules the value is between 330 and 460 µm.

The dimensions X_{M2} and Y_{M2} of the encapsulation depend strongly on the chip module technology used and the size of the assigned chip.

The thicknesses d_T and d_G are the same for chip modules with six contact areas as for chip modules with eight contacts. Naturally, the punching values X_{M1} and Y_{M1} are smaller than for chip modules with eight contact areas as the assigned chips are smaller.

The importance of selection of the correct dimensions for the chip modules is illustrated by the following examples:

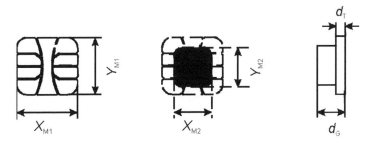

Figure 6.16 Dimensions of a chip module with eight contacts

If the punching measure (Y_{M1}) of a chip module with eight contacts was larger, as mentioned previously , than the chip module would enlarge into the area of the embossing and would be damaged when punching the identification number (see Figure 6.17).

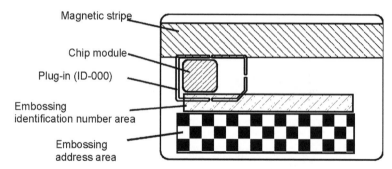

Magnetic stripe

Chip module

Plug-in (ID-000)

Embossing
identification number area

Embossing
address area

Figure 6.17 Areas on the smart card which are reserved for card items other than the chip module

Also with the application for GSM (Global System for Mobile Communication) as plug-in SIM (ID-000) (see Figure 6.17) an excessive chip module could lead to the fact that when punching the plug-in SIM the chip module will get damaged. Of course the ID-000 can only be used if the smart card neither has a magnetic stripe nor an embossing area; otherwise it would be damaged when punching.

If the chip module projects into the area of the magnetic stripe then the rear side of the card material can easily warp itself when the chip module is placed into the cavity of the card body because of the following embedding by heat and force. Thus the reading characteristics of the magnetic stripe are impaired, which leads to unreliable data communication.

Chip modules with a total thickness $d_G > 620$ µm allow the remaining wall thickness of the remaining card material to shrink under 140 µm (also see Section 4.2.1).

With this small remaining wall thickness the fact can occur that the chip module can be seen from the rear side. This happens if 100 µm of a transparent overlay foil is used on the rear side of the smart card when only 40 µm from the foil is remaining, which is not sufficient for perfect cover.

The described materials are standard at present and are being constantly developed further and optimized. For example, there is some thinking about replacing the substrate and the use of polycarbonate foils in place of FR4, which could possibly lead to cost savings during the chip module production.

6.4 Production processes for chip modules

The preceding section demonstrated the many different materials available which open up a multiplicity of combination options. The manufacturing methods are just as various. Here two different exemplary production methods are described.

In the first example a contact chip module with a stiffening frame and an epoxy resin glob top is described and subsequently a contact chip module with an encapsulation from thermosetting polymers (injection-moulding technique) is depicted.

Figure 6.18 represents the necessary individual components of a chip module for this example. The heart of a chip module is the chip. Its method of manufacture is described in Section 5.2.

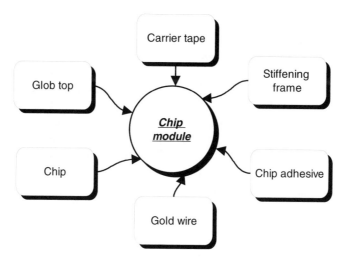

Figure 6.18 Required individual components for a chip module with stiffening frames and epoxy resin encapsulation

The next item which is needed for the chip module is the carrier tape which is manufactured as shown in the flow chart in Figure 6.19.

As raw material FR4 (epoxy resin, glass-fibre reinforced) with a thickness of 110 µm is unwound off a broad roll.

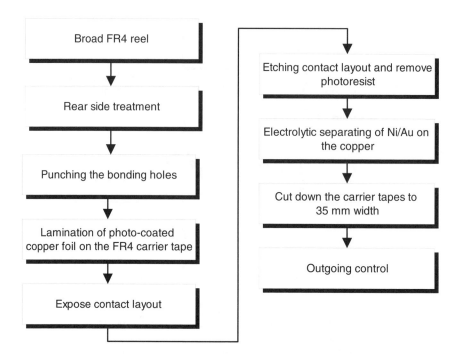

Figure 6.19 Flow chart for the production of FR4 carrier tapes

In the following work procedure the rear side of the carrier tapes is roughened up by sandblasting or with wire brushes. Figure 6.20 shows the scanning electron microscope (SEM) picture of a carrier tape surface and how it is required for the connection of the chip module using heat-activatable adhesive with the card body. If a cyanoacrylate adhesive is used for the connection technique, roughening the carrier tape up can be omitted. Figure 6.21 shows a carrier tape surface without grinding. The difference is present in the surface texture. With the roughened up carrier tape, tiny particles are ripped out from the carrier tape whereby microscopically small sharp-edged craters develop in which the heat-activatable adhesive can tooth itself. Without surface treatment the carrier tape surface is very smooth and for heat-activatable adhesives only of limited suitability.

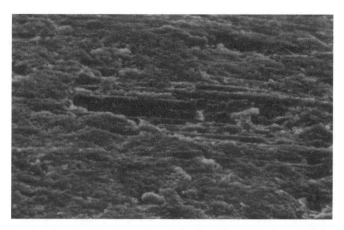

Figure 6.20 SEM accommodation of a sandblasted surface, which is used for heat-activatable adhesives (x750 magnification)

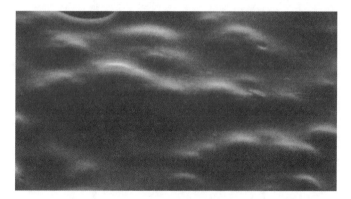

Figure 6.21 SEM accommodation of an untreated surface, which is used for reactive adhesive systems (x750 magnification)

After surface treatment the holes for the later connection between the chip and chip module contact and a window for the chip are punched out (see Figure 6.22).

Figure 6.22 Window from a carrier tape with free punching for the chip and wire bonding to the contact areas on the front. Additionally, punching at the edge for the perforation holes

Additionally, perforation holes with a size of 1.42 mm * 1.42 mm and a separation distance of 4.25 mm according to the JEDEC S 35-Standard are punched. The perforation holes are used throughout all the following production steps as points of reference and as sprocket holes (see Figure 6.22 and Figure 6.23).

As a next manufacturing step, a photo-coated copper foil, which indicates an adhesive layer, is up-laminated on the punched carrier tape. The copper foil covers the punching holes. Afterwards, the contact layout is applied on the copper-coated carrier tape by a photomask. Using the following etching process the outline of the later contact areas is free-etched.

The free-etched copper contact areas still have to be improved with a protective layer of nickel and gold. This step is necessary as the copper surfaces would oxidize quickly and the electrical functionality would no longer be ensured. In addition, copper is a soft metal and would not withstand mechanical loads in everyday use. The nickel layer also prevents copper diffusing into the gold and vice versa.

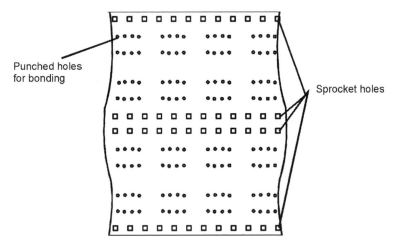

Figure 6.23 FR4 carrier tape with punched holes for bonding and sprocket holes according to JEDEC S 35

In order to improve the copper contacts, a thin nickel/gold layer is applied on the contact areas by electrolytic separation. Additionally, the copper surface over the punched holes is also improved. This is necessary for the wire-bonding process to enable good adhesion of the bonding wires.

Now the broad FR4-roll is adapted for 35 mm broad carrier tapes (see Figure 6.24). Afterwards the carrier tapes are wound up on a transportation reel.

Now the carrier tape and the chip are ready to be united in the chip module manufacturing process.

For the example of a chip module with a stiffening frame and a sealing compound hardened with UV radiation the further production steps are summarized in the flow chart shown in Figure 6.25.

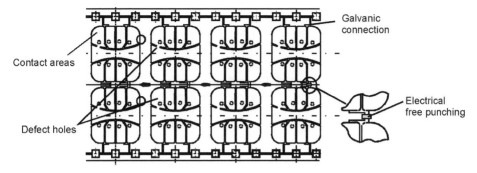

Figure 6.24 Module tape after free etching of the contact areas and cut from the broad starting tape

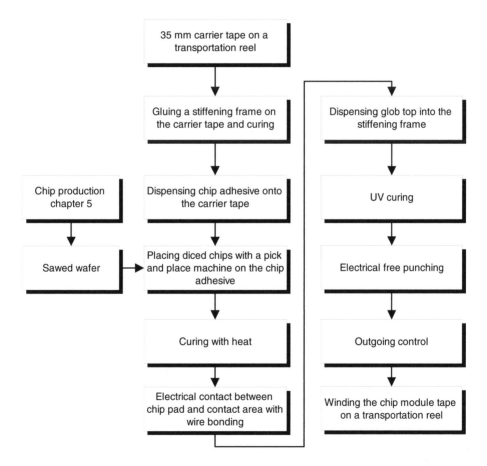

Figure 6.25 Flow chart for the production of a chip module with stiffening frame and UV-hardened sealing compound

First the stiffening frame is glued on the carrier tape and then hardened thermally. The size and position of the stiffening frame depends on the position of the punch holes for the later wire bonding. The stiffening frame should not be too close to the punch holes, otherwise the bonding head pushes into the stiffening frame and may get damaged. Afterwards, the position of the stiffening frame is checked with an automatic visual system.

Next the chip adhesive is spread on the carrier tape. Depending upon chip size one or more portions of adhesive are applied. This can be implemented by a dosing system where each point of adhesive is dispensed individually on the carrier tape.

A further possibility for transferring adhesives on the carrier tape is to dip a stamp into a small pot which is filled with adhesive to a certain thickness (see Figure 6.26). This process is called dipping.

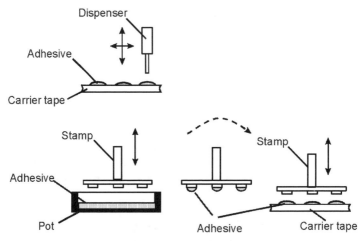

Figure 6.26 Two procedures for the dosage of chip adhesive points on the carrier tape (dispensing at the top; dipping at the bottom)

When pulling the stamp out of the small pot adhesive droplets remain hanging on the stamp. Then the stamp is placed on the carrier tape and transfers the adhesive droplets to the carrier tape. When organizing the position of the adhesive points one should ensure that the corners of the chip are completely underfilled. Then the chip cannot tilt during the later applying (die attachment).

The chip is removed from the adhesive foil with a so-called 'pick and place' machine and placed on the still-liquid points of adhesive.

The adhesive foil is stretched in a frame. A clamp presses it against a locking ring (clamping ring) located under the adhesive foil (see Figure 6.27). This makes is easier to remove the chip from the adhesive foil.

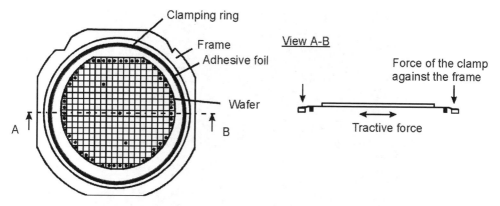

Figure 6.27 Wafer on clamping frames in die-bonder

Thus the adhesive foil becomes strongly strained, and the individual chips are separated by extension of the sawing streets in the sawed wafer. To separate the chip from the adhesive foil, it is raised from the back side using small needles in order to overcome the adhesion of the adhesive coating of the foil. At the same time the raised chip is removed using opposite suction tweezers (see Figure 6.28). Depending on the size of the chip, one or more needles are necessary for extraction from the adhesive foil.

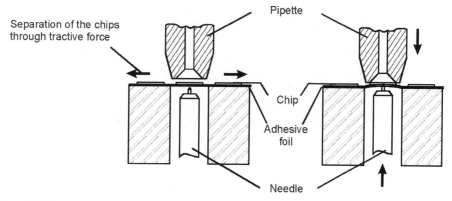

Figure 6.28 Removal of a chip from the adhesive foil with support from needles and suction tweezers in a pick and place machine

Figure 6.29 shows a die-bonding machine.

Subsequently, the chip is positioned on the points of adhesive located on the carrier tape. After placing the chip the adhesive thickness is approximately 25 µm and the adhesive should be full-laminar under the chip. This is ensured if some chip adhesive appears underneath the chip but is should not climb up over one-third of the chip edge. If the chip adhesive rises over the chip edge it may cause short circuits on the chip surface.

Figure 6.29 Die-placer picks a chip from a wafer and places the chip on the carrier tape (source: Mühlbauer)

In order to avoid the positioning of defective chips, a check is made on whether the chips are marked before they are taken from the adhesive foil. The marking of defective chips takes place after the electrical wafer outgoing inspection. Defective chips are marked with a black ink dot. Also a so-called 'wafer map' is common: the positions of the defective chips on the wafer are stored as a data file on floppy disk. The die-bonder detects these markings and does not take the chips off the adhesive foil.

When removing the chip from the adhesive foil it becomes visible, provided the sawn wafer has been well sawed. If the sawing blade cuts too deeply into the adhesive foil it could tear by extending. In this case the adhesive has to be agglutinated again, which involves a lot of work. Another critical problem occurs when larger silicon fragments splinter off the rear side of the chips and lie in the sawing street. The fragments can be carried forward on the remaining wafer, when the chip is picked from the adhesive foil. In the worst case these fragments could damage the passivation layer of the remaining chips.

Also storage of the wafer on the adhesive foil beyond the permitted storage time is critical. The adhesion of the adhesive foil increases with ageing, and it becomes more difficult to loosen the chip from the adhesive foil. In the extreme case this leads to the fact that the chip cannot be removed any longer with a pick and place machine from the adhesive foil or that the chip drops off the suction tweezers.

In the next production step the carrier tape with the placed chip is led through a heat tunnel, whereby the chip adhesive is hardened. Figure 6.30 shows a segment of a carrier tape with a mounted chip.

Now the electrical connections between the chip pads and the contact areas of the carrier tape are made with a wire bonder. The wire used has a diameter of 25 µm.

Figure 6.30 Segment of a carrier tape with a mounted chip before the chip is connected with the contacts of the carrier tape

The mostly widely used bond process is thermosonic welding with ball—wedge bonding machines, which uses gold wire. Also from time to time ultrasonic welding with wedge-bonding machines is used. This uses an aluminium wire.

Thermosonic welding connects the electrical pads of the chip with the chip module contact areas through a welding bond formed by ultrasonic power, pressure and heat. Figure 6.31 represents the sequence of a ball—wedge bond process.

Here the gold wire is led through a capillary and melted underneath it by the energy released from a spark discharge. The melted wire forms a ball due to its surface tension. Next the capillary is lowered onto the chip pad and with pressure, warming and ultrasonic agitation forms the welded joint between the gold (Au) wire and the aluminium (Al) pad. A nail-head-shaped connection between the bond wire and pad is established. The form of the connection is given by the shape of the capillary opening, the so-called nail head. Now the capillary is raised and pulled over the contact area of the carrier tape. Due to the symmetry of ball bonds the gold wire can be pulled in each direction. That is an important advantage in comparison to the wedge—wedge process with use of an aluminium wire, where this is not possible.

When setting down the bond wire on the improved gold contact surface of the carrier tape a so-called 'loop' forms. At the contact area a welded joint between the bond wire and gold contact surface is produced by the influence of pressure, heat and ultrasonic agitation.

This contact is in contrast to the first bond where the edge of the capillary is used as a bond wedge. Thereby a break section in the bond wire develops where the bond wire tears off when lifting the capillary up; a so-called wedge bonding results. In this way the electrical connection between the chip pads and the contacts of the carrier tape is established.

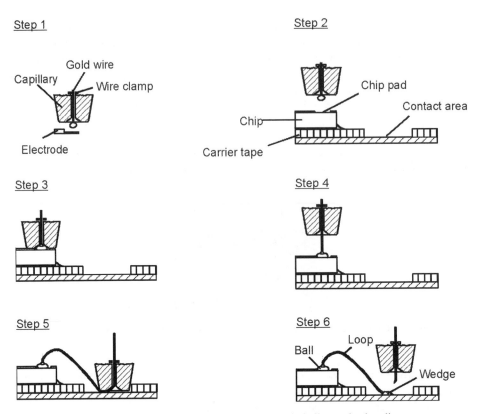

Figure 6.31 Schematic operational sequence in ball—wedge bonding

In order to increase the cycle time in chip module production, two wire bonders are usually used. One machine bonds the first track of the carrier tape, and the second bonder bonds the second track. ESEC and Mühlbauer are two companies which distribute wire bonders for chip modules.

In summary, every chip module is equipped with a stiffening frame and a chip, which is connected with the contacts of the chip module via bond wires. The next production step involves the use of carrier tape.

In order to protect the chip and the bond wires against external influences, encapsulation must take place as a further production step. This is done with a viscous epoxy resin which is dispensed into the stiffening frame.

When dispensing the sealing compound it is vital to ensure that no cavities or air bubbles are formed which may later cause a separation between the chip and carrier tape. Also the bond wires should not be shifted during dispensing by the dosing needles in order to avoid mechanical damage or short circuits between the bond wires.

The dispensing of the epoxy resin should not flow over the frame and contaminate the remaining carrier tape as this surface is needed for the later embedding of the chip module for bonding in the smart card.

After dispensing, curing by UV light takes place. One has to ensure that the exposure time and intensity of the UV lamp are sufficient in order to activate the sealing compound completely. Also the UV lamp should be replaced from time to time as the intensity decreases during its lifespan.

Now the chip and bond wires are protected.

Figure 6.32 shows another widespread possibility, the so-called 'dam and fill' encapsulation for large chips.

Figure 6.32 Dam and fill encapsulation technology for large chips (left: dam dispensing; right: fill up of the dam) (source: DELO)

First a dam is dispensed around the outline of the bond wires. Afterwards the dam is filled up with the sealing compound and then cured. For this thermally hardening epoxy resins are used. With newly developed materials from the DELO company it is also possible to use UV hardening.

With thermally hardening sealing compounds it is important that the curing time and temperature are sufficient for complete hardening. The hardening temperature selected must not be too high in order to protect the carrier tape against thermal harm.

Figure 6.33 is a schematic representation of a encapsulation machine from the Mühlbauer company. The dispensing unit and optical checking devices are in the front section. Then follows the curing section for thermally sealing compounds. For UV sealing compounds the UV lamp is at this position.

Returning to the initial example — now the finished carrier tape with fully cured encapsulation and a stiffening frame is present.

The chip module is now almost finished for the electrical outgoing inspection. Before that the contact areas of the chip module must be separated electrically from each other by so-called free punching.

The electrical free punching is necessary as during the carrier tape production (see previously) all metallic surfaces are connected with each others as a result electrical galvanizing of the copper surface can be executed. Therefore the chip modules cannot be checked electrically after galvanizing as they are on short circuit to each other. The connecting bridges of the contact areas must, as shown in Figure 6.24, become separated with help of a punching tool. This process is called free punching. If the layout of the connecting bridges is designed in such a way that only one or two free punching holes have to be done to separate the contact areas then the design is done well, otherwise there will be a lot of punching holes in the carrier tape and the tape will resemble a 'Swiss Cheese'. This would also reduce the mechanical stability of the carrier tape.

The full electrical functionality of the chip module is checked with the following electrical outgoing control. Defect chip modules are marked by a punched defect hole, as can be seen in Figure 6.24, in order to avoid the unnecessary embedding of defect chip modules into the smart card.

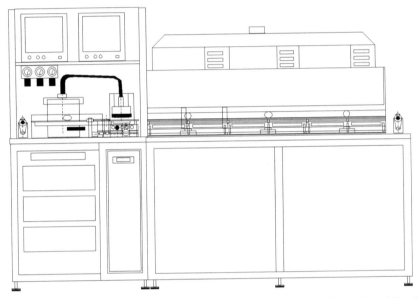

Figure 6.33 Schematic structure of an encapsulation machine to cover chip modules with a following hardening curing unit (source: CME3010/CU1600TUV Mühlbauer)

The position of the defect hole should be chosen in such a way that the electrical functionality of the chip module is not impaired, otherwise it is not possible to undertake electrical analyses later. The defect hole, however, should be situated within the outline of the punched chip module, thus a chip module which is embedded into the smart card body by mistake can be detected by visual outgoing inspection.

After the electrical and visual outgoing inspection the chip module tape is wound on a transportation reel (see Figure 6.34). For protection from contamination and scratching of the chip module surface, an intermediate tape is added as separation between each wound layer.

In selection of the transportation reel the inside diameter should not be too small. Too small inside diameter may cause an extremely tight bending radius and the chip modules become damaged mechanically. In Figure 6.34 an inside diameter of 100 mm was selected.

Now the transportation reel filled with chip modules can be sent away to the smart card manufacturer or if the manufacturer does the packaging in-house then the reel can be passed over to the smart card production line.

For chips with a small surface the encapsulation can be done without a stiffening frame. Therefore only one drop of epoxy resin is dispensed on the chip and bond wires.

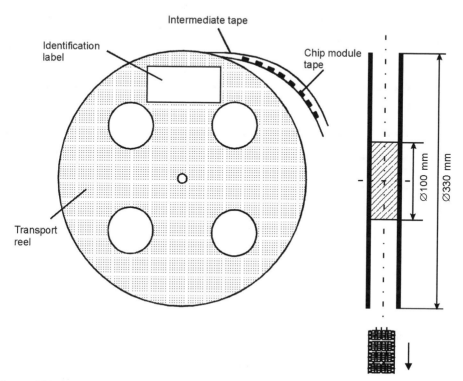

Figure 6.34 Side view and front view of a transportation reel with an example of a wound up chip module tape

As announced at the beginning of this section, we now present a second procedure for the production of chip modules.

This procedure for chip module production is injection moulding. The individual process steps are described in the flow chart shown in Figure 6.35

Here the chip and bond wires, in contrast to the previous example described, are enclosed by thermosetting polymers. The difficulty of the procedure involves the fact that the chip and the lead frame are not completely encapsulated as is the case with dual in line packaging (conventional semiconductor packages for soldering on printed circuit boards). Regarding a chip module for smart cards the encapsulation is only on one side of the carrier tape which make the process much more critical.

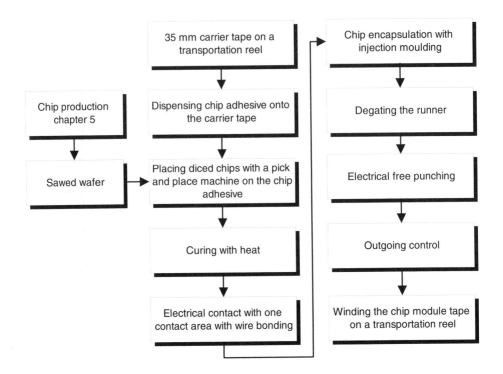

Figure 6.35 Flow chart for the production of a chip module with injection moulded cover

For this the carrier tape with the chip and the bond wires is held between two halves of a mould. Then the injection-moulding compound (thermoplastic resin) is pressed into the two halves of the mould with high closing forces. The difficulty of this process is that the resin does not enclose the carrier tape. The resin is only on the chip side of the carrier tape. When the closing force of the two halves of the mould is too low the moulding compound between the upper halves and the carrier tape may flow out, this is called 'flash'. Flash can have a negative effect in the embedding technique of the chip module in the smart card. Therefore closing force of the moulds must be very high to avoid flash.

With too high a closing force, however, the mould is impressed on the contact area, which is optically not acceptable.

Among other things flash can also result from unevenness in the mould or in the carrier tape. Therefore the mould is usually cleaned with brushes after each moulding process in order to avoid unevenness by contamination of the tool. Naturally the mould tools have several uses. For example, 32 chip modules are moulded in one work procedure.

In contrast to the resin-proportioned chip modules the bonding wire shift is a critical factor with the chip modules manufactured by the injection-moulding technique. This shift is caused by injecting the thermosetting polymer into the mould. The bond wires and their junction points are mechanically strongly loaded and can be blown or torn off. In the most unfavourable case this can lead to short circuits with blowing of the wire. This unfavourable side effect however offers the possibility to find incorrect bonding wire connections reliably since these cannot withstand the mechanical load and will be detected with the electrical outgoing inspection as defective chip modules.

This problem of blowing of wires can be solved when the viscosity of the moulding compound, the injecting rate and the pressing power are coordinated favourably.

Further critical areas are the joints between two carrier tapes which are interconnected by a heatproof capton tape. Joints develop if, for example, parts of the carrier tape are taken out for process control or one carrier tapes ends and a new carrier tape has to be added to the end. Joints are extremely problematic for the injection-moulding process as the moulds cannot completely close and cause increased flash spread out.

In order to bring the moulding compound to the chips and bond wires, channels between the chip modules are needed (see Figure 6.36) — so-called runners. Here runners are developed which must be removed in a separate manufacturing step after injecting the moulding compound and opening the mould. Since the moulding compound adheres extremely strongly on the carrier tape, the carrier tape surface would be heavily damaged when taking the runners off.

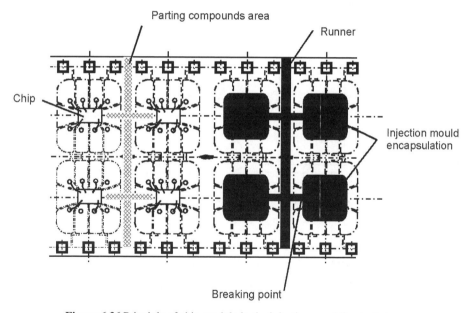

Figure 6.36 Principle of chip module in the injection-moulding technique

Therefore within the area of the runners parting compounds are applied. These parting compounds may be chemical or gilded metallic surfaces, for example. With separating the runner, the transition between runner and encapsulation must be arranged in such a way for the edge of the chip module that a break section develops. Here the runners are defined broken at the edge of the chip module. In addition, there are already solutions where no more runners on the carrier tape are necessary. Therefore the moulding compound is injected directly over the chip module.

The following production steps, e.g. free-stamping machines for the contacts, outgoing inspection and winding up of the chip module tape on a transportation reel, are again the same as with the chip module with stiffening frame and UV-hardening glob top.

The better technique for production of chip modules depends strongly on the volumes which have to be produced, the constantly changing chip sizes and the necessary flexibility of manufacturing.

There is a set of quality-assurance measurements which help to ensure the quality of the final product in the chip module production. Starting with simple checks such as carrier tape thickness, roughness of the carrier tape surface, geometry of the contact areas, total module thickness and the general optical appearance of the contact areas up to the more complicated examination of the bonds.

The bonding wire connection is one of the critical constituents of the chip module. Figure 6.37 shows the critical areas of a ball — wedge connection.

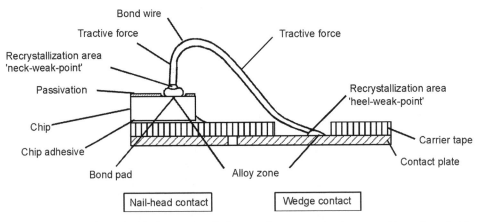

Figure 6.37 Overview of critical areas in a ball — wedge connection between chip pads and chip module contacts

The criteria for a good bonding connection are described in MIL-STD 883 (military standard). Here, we present some selected optical criteria:

- Diameter of the ball contact not smaller than double and not larger than sixfold wire diameter.
- Fifty per cent of the area of the wire contact to be situated within the chip pads
- Wires may not over-cross other bond wires or bond connections
- Development of intermetallic phases

In order to check the bond quality and adhesion of the chip to the carrier tape, we have the pull test, the share test and the peel test.

With the pull test, as shown in Figure 6.38, the gold wire is raised and pulled upward with a small hook until it tears off. The strength needed is measured and gives a value for the stability of the connection between the gold wire, the chip pads and the contact area. The weak point is at the transition wedge bond (heel-weak-point) to the undeformed wire and at the beginning of the wire over the ball bond (neck-weak-point). This test is especially suitable for examination of the bond machine work parameters. A typical value for a good bond is at least 3 cN.

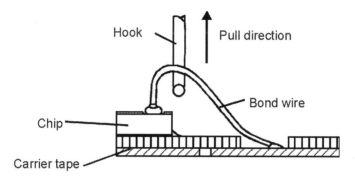

Figure 6.38 Schematic structure of a wire-pull test device

The share test serves to check the adhesive strength of the chip on the carrier tape (see Figure 6.39). For this a small chisel is placed on the side of the chip, which tries to cut the chip off the chip adhesive. With a good connection the chip should break, but the adhesive should not split apart.

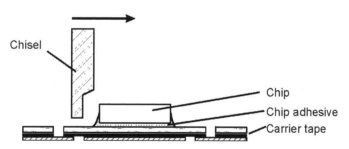

Figure 6.39 Schematic structure of a share test device

The share test can also be used for ball connection. For this, the chisel is set beside the nail head and pressed against the ball connection. The connection is good if the metallization of the chip pad proves itself.

Lastly, we describe the peel test. Before starting the test the wire loop is separated briefly behind one of the contacts and the loose end of the wire is caught in pliers. Afterwards the end of the wire is pulled perpendicular to the chip contact, until the contact becomes detached (peels) or the wire tears. Consequently the maximum needed traction force is considered the measured value.

Each of these tests is executed at regular intervals during chip module production.

Now the chip module is ready for further processing within the production of a smart card, as described in the next chapter.

7 Production processes for smart cards

In this chapter we demonstrate, how the chip modules from Section 6 and the card body from Section 2 are united to form a smart card.

There is a multiplicity of possibilities in order to bring chip modules into the card body and create a smart card.

The selected method of production depends strongly on the respective application of the smart card. With a contactless memory chip card, with which the chip module is inside the smart card, the laminating technique is predominantly used.

With contact smart cards the chip module is positioned in such a way that the electrical contact areas of the chip module can be contacted from the outside using a smart card reader. For the contact smart card there are essentially two procedures for bonding the chip module permanently in the card body: facilitate the connection by means of reactive adhesives (for example, liquid cyanoacrylate adhesives) or with heat-activatable adhesive systems. In this chapter our focus is more on the connection technique with heat-activatable adhesives, the so-called mounting technique. We briefly also refer to the special features or differences of the reactive adhesive systems within the individual production steps.

7.1 Smart cards with contacts

Today for embedding chip modules into the smart card the mounting technique is used predominantly. It involves a procedure with which first a cavity for the chip module is manufactured in the card body. Subsequently, the chip module is provided with a heat-activatable adhesive, brought into the prefabricated cavity under application of force and heat and is connected permanently. Figure 7.1 gives an overview of the production flow during smart card production using the mounting technique.

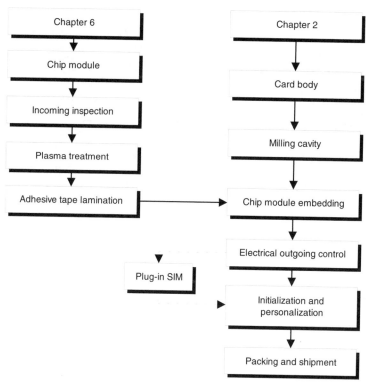

Figure 7.1 Production flow for the manufacture of a smart card using a heat-activatable adhesive

7.1.1 Chip module incoming test

Before smart card production can start, the chip modules delivered on transportation reels are checked in the incoming control machine. Figure 7.2 represents an incoming control machine (or test handler) for chip modules.

First the transportation reel (left side on the figure) is clamped into the machine, transported under the individual workstations and finally wound on a transportation reel at the right side of the machine. With the unwinding of the chip module tape the intermediate protecting tape is coiled on a transport reel; with the winding of the chip module tape on the transport reel an intermediate tape is inserted again.

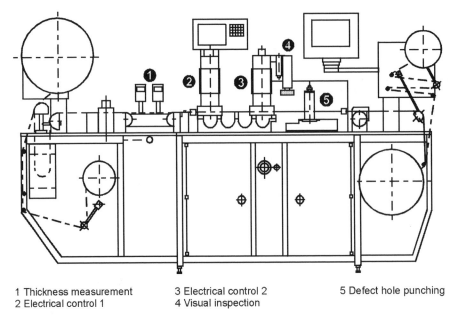

1 Thickness measurement   3 Electrical control 2   5 Defect hole punching
2 Electrical control 1   4 Visual inspection

Figure 7.2 Test handler for chip modules (source: CMT 6520/L Mühlbauer)

With the incoming control machines 100 % of the chip modules can be controlled; however, it is very time-consuming. Therefore usually only samples out of the delivered chip modules are checked.

The range of the check is described by an AQL value (acceptance quality level) (also see Section 9.1). The AQL value and the test range are agreed upon in arrangement with the supplier.

For example if an AQL value of 0.65 is used, this means for the case of a chip module supply of 12,000 functional chip modules 315 chip modules are checked. The chip modules which are provided for checking must be selected in such a way that they are distributed randomly over the entire delivery volume. If four of the 315 checked chip modules do not meet the requirements, then the chip module delivery may be returned to the manufacturer. With three defective chip modules on the reel the delivery as released is approved for further production.

First the thickness measurement takes place in the shown incoming control machine. Here the thickness of the carrier tape is measured. The position of the measuring point should avoid the areas of free punching (defect hole or electrical free punching), since the existing punch side at the edge of the defect hole would falsify the measurement. As probe tip, a ball with a diameter of 2 mm works satisfactorily. In the second thickness measuring station the total thickness of the chip module can be measured. In particul for chip modules for which the glob top is made from liquid epoxy resins not only the specification of a maximum thickness is important, but also a lower thickness limit. The lower thickness limit is important because here chip modules can be out-measured, for which the glob top does not cover the bond wires completely. The wires are exposed to mechanical and chemical influences and can lead to a later failure.

At the next test station the electrical functionality is tested. In order to enable the execution of the electrical test usually spring-suspended golden probe heads are put on the contacts of the chip module. The machine shown can check 32 chip modules at the same time. For reliable contacting of the chip module, the probe head must be put on the contact with low pressure. A needle-shaped probe head is used for the test probes. With probe heads in crown shape, dirt particles can stick more easily and influence reliable contacting.

For the electrical check there are different testing methods. For example, the ATR (answer to reset) of the chip can be checked. A command is transmitted to the chip, which processes a small test program internally and sends a response sequence back to the machine. The chip is working correctly, if this corresponds with the expected sequence. A more intensive and also time-consuming test (besides the ATR test) consists in writing and reading different data samples into the EEPROM. With this test the entire chip is working. Two possible failures in the chip can be tested with the EEPROM test: First, the reliability of the EEPROM and physical damage on the chip surface as the EEPROM of a 4 KB EEPROM covers over 50 % of the complete chip surface. Second, the electrical characteristics of the chip can be measured.

The electrical test station can also be used as a workstation for the initialization. During the initialization of the microcontroller chip a customized program is brought into the chip. For this all chip modules must be initialized.

At the last test station the visual check is implemented in the shown incoming control machine. Here the contacts of the chip module are checked for pollution and scratches using a camera system.

This check is sometimes problematic as detecting visual errors with a camera system in some cases can be weighted differently by chip module manufacturers and smart card manufacturers. Therefore a visual error catalogue, where suppliers and customers have agreed upon the weighting beforehand is very helpful.

As the last workstation a defect hole punching station is integrated in the machine. It marks the chip modules which are detected as defects in the test stations and afterwards punches a defect hole into the carrier tape (see Figure 6.24). In order to facilitate the later traceability of the thus marked chip modules, it would be helpfully if the smart card manufacturer punches a somewhat smaller (e.g. Ø 1.8 mm) defect hole than the chip module manufacturer (e.g. Ø 2.0 mm), or vice versa.

In addition to the electrical, thickness and visual checks the incoming control machine also counts the numbers of the supplied chip modules.

In the end the checked chip module tape is wound on an empty transport reel and is now available for further processing.

7.1.2 Production of the card body cavity

The thus defined card body, prefabricated with various card items (e.g. magnetic stripe and hologram), must have a cavity for embedding of the chip module.

For this there are two possibilities: On the one hand the cavity can be produced by the use of the injection-moulding technique (see Section 3.2). This is the norm with simple memory chip cards, for example telephone cards.

If smart cards with several card items are to be manufactured, for example holograms, pictures and other card items, as described in Section 2.4, a card body is used which is laminated from different plastic foils. The milling technique is used to produce the cavity in such laminated card bodies.

Milling machines with one or two milling spindles are used. Figure 7.3 shows the schematic structure of a milling machine from the Mühlbauer company. Milling of plastic is problematic and requires an exact tuning of the milling head's feed speed and the rotational speed of the milling tool. In order to eliminate the heat developed with end milling, the milling tool is cooled by compressed air. Additionally developing milling splinters are sucked off from the milling area by a suction apparatus. Sucking off is important, ensuring that milling splinters do not get under the card body which is milled. This could cause false milling depths and imprints on the card's back side. Milling splinters which possibly remain in the cavity geometry impair the later adjustment of the chip module in the cavity.

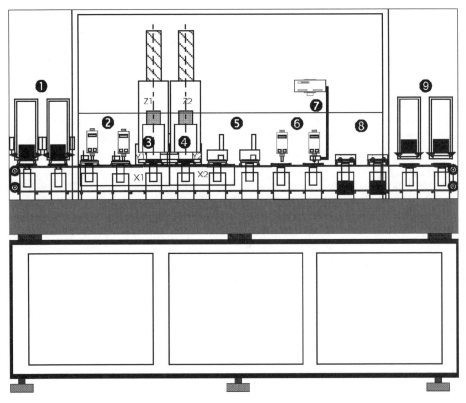

1 Card magazine input	4 Milling spindle 2	7 Cavity depth measurement 2
2 Thickness measurement with	5 Cleaning station	8 Reject/sample station
card orientation verification	6 Cavity depth measurement 1	9 Card magazine output
3 Milling spindle 1		

Figure 7.3 Schematic structure of a milling machine for the milling of a cavity in a card body (source: SCM 5030/6C0 Mühlbauer)

In the next workstation the remaining milling splinters are removed out of the cavity. This can be done with rotary brushes or with a specific rotating suction device.

After this follows the thickness measurement station which checks the depth of the cavity. If the actual dimensions, with consideration of the tolerances, deviate from the nominal dimensions, the depth control of the milling machine regulates the milling spindles independently up or down.

The reject and sample output station follows as the last workstation. In this station card bodies which are too thick or were inserted falsely into the milling machine are removed. There is also the possibility of taking milled card bodies off the milling machine for an external dimension measurement without disturbing the production run. This enables manufacture-accompanying quality controls.

Additionally the milling machine as the first workstation has a card body thickness measurement station with integrated card orientation recognition. The orientation recognition system using optical sensors detects light bright/dark areas on the card body and hence can determine whether the card body was inserted into the milling machine with the correct side and in the correct orientation. For this purpose there are marks printed on the card surface within the future cavity area which can be detected by the machine.

Independently of the milling machine the dimensions of the cavity in the card body are looked at. For this, different solutions are pursued.

The position of the chip module contacts is defined in ISO 7816. From this the position of the chip module in the smart card (see Figure 7.4) can be determined. For a chip module with eight electrical contacts a central point results in the case of a symmetrical module design, from 15.06 mm in the X-direction and 23.89 mm in the Y-direction with reference to the upper card edge. For a chip module with six electrical contacts the Y-value is 22.62 mm; the X-value remains the same. Despite the parameters given from the ISO specification there are a multiplicity of parameters which can be varied in order to influence the functionality of the smart card.

The milling geometry which is produced can be divided into two levels: The first level is the adhesive surface z_1, on which the chip module is later glued into the card body. The second level z_2 is the remaining wall thickness of the smart card in the cavity. Figure 7.5 shows a double milling station. Using this the first level could be milled with the first milling spindle and the second level with the second milling spindle. But this is not very favourable as the tolerances of the individual spindles would add themselves together over the two levels in the worst case. The best method is as implemented in Figure 7.3 for each milling spindle to mill the complete cavity. Here only the small milling tolerance of one milling spindle has to be considered.

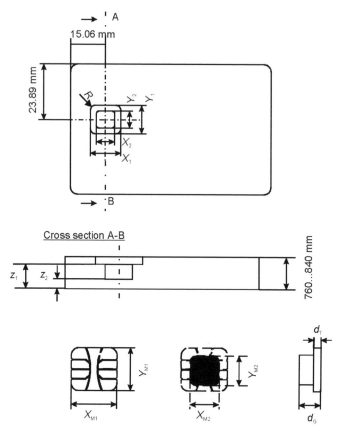

Figure 7.4 Card body with cavity for a chip module with eight contacts and a chip module with essential dimensions

During the sizing of geometry for the adhesive surface z_1 the chip module's external dimensions (X_{M1}, Y_{M1}) plus a circulating gap (α, β) around the chip module are estimated. Using Equation (7.1) the dimensions (X_1, X_2, Y_1, Y_2) can be calculated for the cavity geometry.

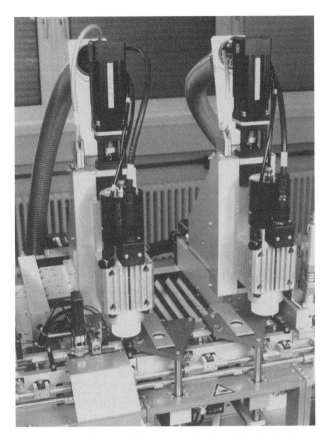

Figure 7.5 Figure of a double milling head for the production of the cavity in a card body (source: Mühlbauer)

The gap is used to compensate the tolerances which occur during milling and the placing of the chip module into the cavity with the later embedding process. The gap is also important in order to reduce the probability of placing the chip module outside of the cavity. The consequence of this would be that the chip module stands out from the smart card surface leading to scratches on the back side of the upper smart card when stacking the finished smart cards in the feed magazine. Additionally, perfect adhesion of the chip module in the smart card is not completely ensured.

$$X_1 = X_{M1} + \alpha; X_2 = X_{M2} + \alpha$$
$$Y_1 = Y_{M1} + \beta; Y_2 = Y_{M2} + \beta \tag{7.1}$$

If α and β are selected too large then the general optical impression of the smart card is impaired by the circulating white gap around the chip module.

In order to minimize the circulating gap and increase the adhesive of the chip module in the smart card, occasionally a milling tool with a negative angle is used. The adhesion is increased by the fact that the heat-activatable adhesive withdraws into the formed cavity and leads to an additional adjustment of the chip module at the edge (see Figure 7.6). This increase of the connection force has a positive effect on the physical test required according to ISO 7816.

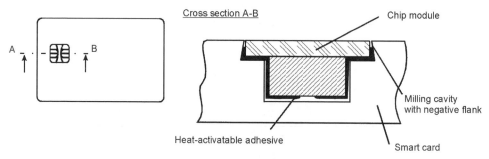

Figure 7.6 Milling cavity with negative flank

Further important dimensions for the definition of cavity geometry are the depths in the z-direction. z_1 is relevant for the adhesive surface and z_2 for the remaining wall thickness. The depth z_1 predominantly depends on the thickness of the chip module carrier and the embedding technique used for the smart card. If a maximum smart card thickness of 840 µm is assumed according to ISO 7816 and a chip module carrier tape of typical 160 µm with an adhesive layer of 40 µm, the adhesive surface should be with 640 µm (840 µm – 160 µm – 40 µm = 640 µm).

If the general machine and material tolerances are considered, the measure of z_1 has a value of approximately 600 µm. The ISO 7816 allows a protrusion of the chip module over the smart card surface of 50 µm maximum; however, the total thickness of the smart card cannot exceed 840 µm. A protrusion of the chip module however leads in most cases to the fact that when later stacking the smart cards into a feed magazine the smart card lying above is scratched in each case.

The depth z_2 of the remaining wall thickness is determined by the total thickness of the chip module d_G. This value is somewhat more difficult to determine and depends a great deal on the used embedding technique. With heat-activatable adhesive a small gap is left over between the chip module rear side and the remaining thickness z_2 of the smart card.

The gap also provides mechanical decoupling of the chip module within the critical area of the chip, which reduces the danger of a chip break.

When using heat-activatable adhesive the remaining wall thickness z_2 of the smart card should not be too thin. With a polyvinyl chloride (PVC) smart card the depth z_2 should be larger than 160 µm, since during embedding the chip module into the smart card the influencing amount of heat leads to a warping of the smart card rear side.

With the numerical examples mentioned above it can be assumed that thickness measurement takes place from the card back. There is, however, another second possibility of indicating and measuring the thickness, i.e. from the card front side.

The mechanical implementation of the measurement system is the same with both procedures, only the zero point (point of reference) is defined differently (see Figure 7.7).

With the first principle the zero point is on the card surface. This means that each card has to be measured for thickness before the milling process. With the second measurement principle the zero point is on the base plate. The zero point must be only referenced when adjusting the milling machine to the base plate. If one undertakes the first procedure without the thickness measurement, then the card body must be divided into thickness classes before milling in order to avoid a possible protrusion of the chip module. This distinction is not necessary when using the second procedure, since the depths z_1 and z_2 always have the same distance to the base plate. The advantage lies in the fact that the chip module adhesive surface is always in the same distance to the card body surface and protrusion of the chip module is avoided with large certainty.

More detailed information about the calculation of the cavity can be found in Section 4.2.1.

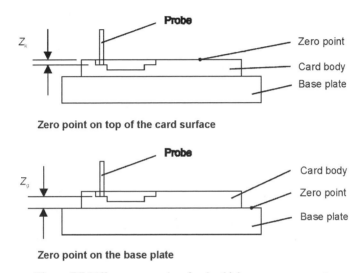

Zero point on top of the card surface

Zero point on the base plate

Figure 7.7 Different zero points for the thickness measurements

If the cavity for the chip module is created in the card body, as can be seen for example in Figure 7.8, the next manufacturing step to finishing the smart card is bringing the chip module into the cavity of the card body and the necessary connection technique for that.

Figure 7.8 Example of cavity geometry for a chip module with eight (left) and six (right) contact areas

7.1.3 Connection technology between chip module and card body

The next manufacturing step in the production of the smart card is the connecting technique between the chip module and the card body.

There is a multiplicity of technical possibilities for connecting a chip module with the smart card. Contact adhesive, reactive adhesives and thermally activatable adhesives are the most frequently used procedures.

Contact adhesives have not become generally accepted as a connecting media between the chip module and card body as the reliability of the connection cannot be ensured over the length of application of a smart card.

As previously mentioned, presently two connection techniques are common. On the one hand there are reactive adhesive systems (for example, cyanoacrylate adhesives), and on the other hand heat-activatable adhesive systems in the form of adhesive foils. Both procedures are found spreading wide in the production of smart cards. In the following, the production of a smart card with heat-activatable adhesive is further illustrated as an example.

To enable use of the heat-activatable adhesive as a connecting vehicle between the chip module and the milled surface in the cavity of the card body, it must be applied on the chip module carrier tape. The adhesive is present as a foil with a thickness of 30 to 80 µm depending on the manufacturer. The heat-activatable adhesive is wound up on a reel and is separated with a silicone layer in a similar manner to double-sided adhesive tape.

In order to increase the adhesion of the heat-activatable adhesive on the chip module carrier tape, the surface of the module carrier tape is pre-treated with a low-pressure plasma. This pre-treatment of the surface is necessary, since with storage and handling of the chip module carrier tape small particles of pollution can settle on the tape surface. In particular, organic substances reduce adhesion between the carrier tape and the heat-activatable adhesive. The surface energy is increased by the plasma pre-treatment and the adhesion is influenced in a positive manner. At the same time organic arrears on the surface of the chip module carrier are removed and hence it gives the surface of the module carrier reproducible properties.

In order to treat the chip modules with plasma, one or more chip module transport reels with the chip module carrier are brought into a vacuum chamber. After evacuation, oxygen (O_2) flows through the vacuum chamber and by means of a microwave generator ($f = 2.45$ GHz) a plasma is produced in the chamber. The high degree of ionization and the associated high concentration of radicals ensure that the contamination is removed — almost 'burned' — completely from the surface. The gaseous waste products developed (carbon dioxide, water vapour and small quantities of ozone) are sucked off from the process chamber.

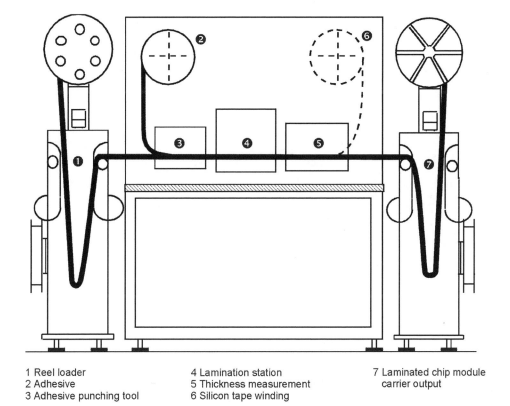

1 Reel loader   4 Lamination station   7 Laminated chip module
2 Adhesive   5 Thickness measurement   carrier output
3 Adhesive punching tool   6 Silicon tape winding

Figure 7.9 Schematic structure of a heat-activatable adhesive (source: CML 3400 Mühlbauer)

With the plasma process the chip module transport reel is only warmed up slightly. The process time takes two to five minutes.

After the plasma treatment the chip module is brought to the next production step, which is laminating of the adhesive to the carrier tape, within a defined time, in order that the surface treatment remains effective.

In the next production step, the heat-activatable adhesive is laminated on the chip module carrier tape before the chip module is inserted into the card body.

Figure 7.9 shows the schematic structure of a manufacturing machine for laminating the adhesive to the chip module carrier tape built by the Mühlbauer company.

During adhesive laminating the heat-activatable adhesive tape, which has the same width as the chip module carrier, less the size of the sprocket holes, is applied on the chip module carrier tape and connected under application of force and temperature with the chip module carrier tape (see Figure 7.10). With 35 mm broad chip module carrier tape a 29 mm broad adhesive tape is used. The heat-activatable adhesive is offered in 50 to 100-m-long rolls by suppliers such as Cardel, TESA and 3M.

As is seen in Figure 7.10, the heat-activatable adhesive tape can have an opening within the area of the chip module encapsulation. This is punched into the adhesive tape of the punching station of the laminating machine. This opening ensures that the chip module does not receive a connection with the card body during the later embedding into the cavity of the card body.

This prevention can beneficially affect the breaking behaviour of the chip module, since no mechanical loads are linked in the area of the chip.

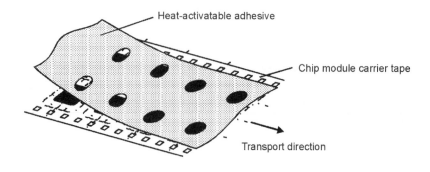

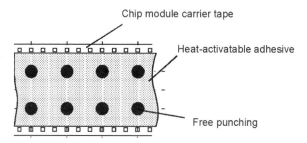

Figure 7.10 Adhesive laminating of a heat-activatable adhesive on a chip module carrier form

The temperatures used by the laminating tools depend on the assigned heat-activatable adhesive. The temperature should be selected in such a way that after laminating the adhesive tape can only be removed from the chip module carrier tape under energy expenditure. Naturally the temperature also should not be selected too high, otherwise the heat-activatable adhesive flows through the defect holes and sticks to the lamination tools. The value for the process temperature is normally situated between 90 and 150 °C.

Figure 7.11 Lamination machine for heat-activatable adhesives (source: Mühlbauer)

Often after the adhesive tape is laminated to the chip module carrier tape with a hot pressing tool another pressing tool follows which is cold. In this step the adhesive is cooled down and fixed for a second time. It is important that the silicon tape (protection tape) remains on the heat-activatable adhesive foil when the adhesive is under the heated station, otherwise the adhesive will stick on the tool and not on the chip module carrier tape.

After fixing the adhesive to the chip module carrier tape there is an outgoing thickness measurement measuring the thickness of the chip module carrier plus the thickness of the adhesive tape. The measurement takes place on the chip module carrier and the adhesive tape. The protection tape of the adhesive should still not be taken off at this stage as it protects the adhesive surface against contamination and prevents possible sticking of the superimposed chip modules on the transportation reel. Figure 7.11 shows a laminating machine used by the Mühlbauer company. With reactive adhesive systems the process step of adhesive laminating is omitted, as this is spread directly into the cavity of the card body. After applying the thermally activatable adhesive, embedding of the chip module into the cavity follows as the next manufacturing step.

7.1.4   Embedding of the chip module

The last production step, before the chip module and the card body come together as a smart card, is bonding of the chip module into the cavity of the card body, this is called implantation.

Depending upon the then used connection technique the flow of implantation of the chip module into the smart card is different.

When using reactive adhesives predominantly liquid cyanoacrylates are used. Points or lines of adhesive are applied on the first adhesive surface z_1 of the smart card cavity. Subsequently, a drop of viscous adhesives is placed into the centre of the cavity z_2. Afterwards the punched-out chip module is inserted into the cavity and fixed with a stamp. The viscous adhesive in the cavity z_2 fixes the chip module in its position, as the final adhesion of the cyanoacrylate adhesive begins. Afterwards the reactive adhesive must be left to harden for a certain time.

Returning to the example of the connection technique with heat-activatable adhesives, the flow for this type of chip module embedding is represented in Figure 7.12.

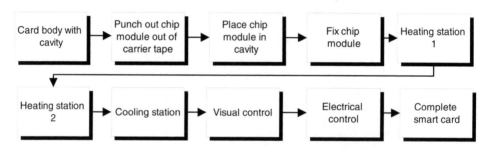

Figure 7.12 Process steps with chip module embedding using heat-activatable adhesive as connection technique

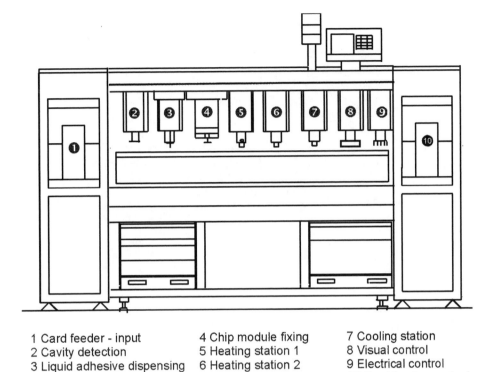

1 Card feeder - input   4 Chip module fixing   7 Cooling station
2 Cavity detection   5 Heating station 1   8 Visual control
3 Liquid adhesive dispensing   6 Heating station 2   9 Electrical control
                                                             10 Card feeder - output

Figure 7.13 Principle construction of a machine for embedding the chip module into the cavity of the smart card (source: SCI 8200 Mühlbauer)

As can be seen in the schematic operational sequence of an implantation machine in Figure 7.13, the card body with a cavity is supplied to the individual process stations outgoing from the card feed unit. First a sensor checks whether a cavity exists in the card body or not.

The next workstation shown in the figure is a dispensing system for reactive adhesives, which is not needed for heat-activatable adhesive systems.

After this the chip module is out-punched from the chip module carrier tape. The chip module punching tool is on the rear side of the implantation machine as shown in Figure 7.14.

Before punching out the chip module from the carrier tape the protecting tape of the heat-activatable adhesive has to be removed, otherwise the chip module would not stick to the card body as the adhesive cannot come in contact with the card material in the cavity (for this see Section 7.1.3).

Only those chip modules are punched out which have no defect hole. The chip module is punched out from the side of the encapsulation to avoid forming of a burr at the chip module edge. In order to avoid damage to the chip module, particularly breaking the chip, the plane of the punching tool is not flat — it must have a cavity at least in the same size as the encapsulation of the chip module.

Figure 7.14 High-speed embedding machine. On the left side of the figure the chip module punching tool can be seen (source: Mühlbauer)

After punching out from the chip module carrier tape the chip module is taken by a suction pipette and transported forward to the transportation system of the implantation machine, in which the card body with the cavity awaits the chip module. The chip module is inserted into the cavity of the card body. In order to prevent jumping out of the still-loose chip module in the cavity with further transport of the smart card, the chip module is attached briefly at the top edge by a heated metal stamp (see Figure 7.15).

Subsequently, with a further heated welding stamp, the same size as the chip module, the connection between the chip module and card body is made. Depending upon the chosen heat-activatable adhesive temperatures from 180 to 225 °C are used. The respective process time and process contact pressure with which the chip module is pressed into the cavity again depend on the used heat-activatable adhesive. A process time between 750 and 1300 ms and process contact pressure between 30 and 150 N is usual.

On heating up the adhesive it is softened for a short time; therefore it is necessary to coordinate the process parameters of temperature, time and force accurately. If the temperature was selected too high or the process time too long, the heat-activatable adhesive may step to the card surface at the edges of the chip module and cause contamination. Additionally, the card rear could be strongly deformed by temperature influence with an unfavourably selected combination of the process parameters. To prevent deformation of the card rear, the bearing surfaces of the individual workstations are often cooled down to between 8 and 15 °C. In order to reduce the process temperature and process time, the heat feeding is divided between two or more welding stations. The reduction temperature not only serves the optics of the smart card rear side, but also helps to reduce the thermal peak loads on the chip module.

Figure 7.15 Transfer of the chip module from the chip module punching tool into the bonding station of the implanting machine and into the cavity of the card body (source: Mühlbauer)

Too high temperature loads can cause a so-called popcorn effect which damages the chip module. Chip modules with epoxy resin sealing compounds are mostly affected by this problem. The popcorn effect is the spontaneous evaporation of the concentrated humidity caused under the thermal effect in the chip module, which can lead to a break of the bonding wire (also see Section 6.3). In order to prevent this effect, the welding stamps should have a cavity within the area of the encapsulation. The temperature is only needed within the area of the adhesive surface, in any case.

The correct mixture of the process parameters also determines the later smart card quality in relation to the adhesion of the chip module in the smart card (also see Section 9.3). Next the smart card continues on to a cooling stamp with ambient temperature or forced cooling.

Here the chip module is fixed and cooled down with a cooling stamp in order that a permanent and at the same time flexible connection between the chip module and the smart card can be made. Now the finished smart card fulfils the mechanical demands according to the usual standards. With reactive adhesive systems a certain time has to elapse before the smart card develops its full mechanical stability.

After the cooling station another electrical and optical short check is made. The optical check takes place via an automatic camera system and withgoing analysis software. Now the smart card is fully functional and can be delivered to the customer, or can be initialized and personalized (see Section 10).

For application within the GSM area the predominant part of the assigned smart cards has no ID1 format, but has the ID-000 format (see Figure 7.16). This small smart card is also termed a 'plug-in SIM'.

The plug-in SIM can be produced by punching it out of the large ID1-smart card. To ensure that the plug-in SIM does not drop out of the ID1 card, it is connected to the ID1 card via bars. In order to facilitate the later out-breaking for the consumer, the bars are weakened by two notches from the front and rear side.

Plug-in SIM punching is undertaken in two steps depending upon the versions. In the first step the plug-in SIM is free punched, and only the bars are left over. In the second step the bars are cut with knifes from above and beneath. After cutting only a very thin part of the bars remains, which develops a breaking point in that section and simplifies the later out-breaking of the ID1.

The position of the bars is not determined by standards and can be arranged following one's own conceptions.

The difficulty in punching out the plug-in SIM lies in the fact that each plastic material behaves differently; therefore the punching tool must be adapted to the card material otherwise the format of the plug-in SIM cannot be kept exactly as described in the ISO. Additionally, the remaining cut-off of the bars should have smooth edges on the plug-in SIM to avoid a later tilting when insert the plug-in SIM into a mobile phone. Also the danger of injuring the user when breaking out the plug-in SIM from the ID1 card should be considered.

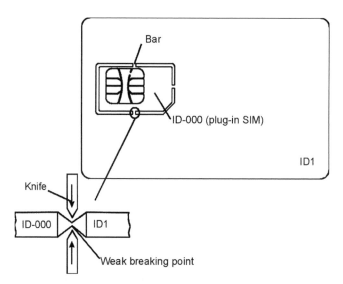

Figure 7.16 ID1-smart card; plug-in SIM ID-000 punched out with knife cutting technique

Normally the plug-in SIM can be put back into the ID1-smart card in a shoddy manner only. There are, however, some solutions to allow this. For example, at three points at the edge of punching in the ID1 card body the card material is deformed easily into the punching area; thus the plug-in SIM can be pressed back into the card again.

After punching for the plug-in SIM is done the GSM smart card can be passed on to the initialization and personalization stage. Following the description of production of a contact smart card using the example of the mounting technique the production of the contactless smart card is demonstrated in the following section.

7.2 Contactless smart cards

The laminating technique was used in the early years of the smart card for the production of contact smart cards. A TAB module was in-laminated into differently punched-out PVC foils.

Lamination refers to construction by placing layers of different plastic foils upon previous layers, which are compressed under effect of pressure and temperature over a certain time. The result is a single plastic sheet. More details are given in Section 3.1.1.

Today, as described in Section 7.1, contact smart cards are predominantly manufactured using the mounting technique and the pure card body is in most cases laminated.

With the introduction of the contactless smart card the laminating technique has experienced a revival for the production of complete smart cards.

A substantial component of the contactless card is the antenna inside the card body which serves energy and data communication.

As can be seen in Figure 7.17 there is a carrier foil beneath the antenna which is connected using a suitable connection technique to a contactless chip module (see Section 6.2), whereupon a foil free-punched opening within the chip module area is situated.

In order to prevent unevenness on the overlay foil a filling material is spread in the remaining cavity around the contactless chip module. This filling up prevents the overlay foils which are applied afterwards from breaking in so the area of the contactless chip module after the laminating process. Hence the card surface is formed smooth and flat. This is particularly important if on the overlay foil are further card items present over the contactless chip module, e.g. a magnetic stripe or a later applied picture, made with thermal transfer printing. With an uneven smart card surface the functionality of the respective card item can be influenced.

In the following sections some different types of antennas and various connection techniques between the chip module and antenna will be presented.

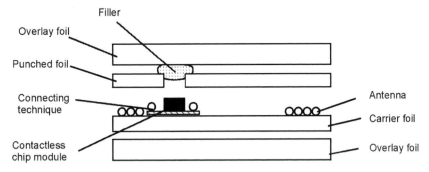

Figure 7.17 Principle construction of a contactless smart card made of different layers of plastic foils

7.2.1 Different types of antennas for contactless smart cards

There are different possibilities for manufacturing an antenna for the contactless smart card.

The antennas must correspond to the electrical requirements of the used contactless systems, for example Mifare® of the Philips company, in each case. Each contactless system has certain requirements for the resistance value, the capacity and inductance of the antenna. From this a certain working range can be obtained. A range of 10 cm between the contactless smart card and the reader is usual with many contactless systems.

The position of the antenna and chip module in the smart card is not determined by any standard. The size and position of the antenna result, however, from the electrical characteristics required by the contactless system. The position of the contactless chip module is strongly limited however by some card items already present on the smart card. Figure 7.18 shows the positions of the most important card items.

Therefore, if for example the contactless smart card is suitable for embossing the antenna cannot cross through a certain area on the smart card otherwise the embossing could lead to an interruption of the antenna and make the contactless smart card functionless. Naturally, also no contactless chip module should be in that area. Also the area below a contact chip module is not a suitable place for the antenna and contactless chip module as they could be milled. Additionally, if possible, the chip module should not be under the magnetic stripe since with unevenness in the card body surface the characteristics of the magnetic stripe could be influenced negatively.

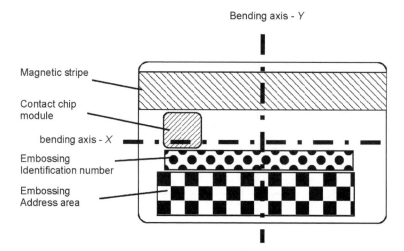

Figure 7.18 Restricted areas for the antenna and the contactless chip module on a smart card

A further critical area for the contactless chip module is situated directly in the bending axis of the smart card. There the mechanical load is the highest and could easily lead to a chip break. For antennas there is a further restriction: If the antenna is too far to the edge of punching of the smart card it could be damaged when punching out the smart card from the sheet.

Here we discuss some types of antenna which are used with the implementation of a contactless card.

First there is the wound antenna. A copper wire coated with insulating varnish is wound on a winding machine around a core, similarly to the production of a transformer.

Afterwards the antenna is placed on a plastic foil and bonded thermally. This type of antenna production is very complex and is rarely used.

An advancement of the wound antenna is the so-called wire embedding technique from the Amatech company. An isolated wire is applied directly on the plastic foil (see Figure 7.19) by means of ultrasonic heating. The copper wire is easily melted into the plastic foil by using ultrasonic energy. This procedure enables very fast production of the antenna and is very flexible regarding the adjustment of antenna geometry at system request.

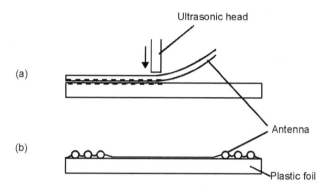

Figure 7.19 Principle of the embedding wire technique for contactless smart cards

A further alternative for the production of the antenna consists of etching the antenna from a copper layer. This is converted as with the wire embedding technique from reel to reel, i.e. the plastic foil does not come in the single-use format, but depending upon the system in 330 mm broad courses with a foil thickness starting from 150 μm.

With the etched antenna conductive lines of at least 100 μm width can be implemented and distances between the conductive lines starting from 100 μm be achieved.

The first production step is a pure copper foil with a thickness between 35 and 70 μm without adhesive is up-laminated on the complete size of the plastic foil using a special connecting technique. Afterwards a photo-sensitive photoresist is applied on the copper foil. The photoresist is exposed through a photomask which contains the structure of the antenna. Subsequently, in chemical baths the clear photoresist is removed and the surplus copper surface is etched away. In the last stage the exposed photoresist is removed and the plastic foil with the present antenna is cleaned thoroughly by chemical additions, dried and rolled up.

A complete removal of all chemicals off the plastic foil is necessary since these could otherwise function as parting compounds between the foils and the contactless smart card and delaminate within this area.

The etched antenna where large numbers of items are involved is a good alternative to the antenna produced using the wired embedding technique.

As a last possibility for manufacturing an antenna we introduce the printed antenna.

The printed antenna, as for the etched antenna, can be manufactured in a continuous procedure reel to reel or in sheets, for example in the format of 310 mm * 510 mm (3 * 8 cards) (see Figure 7.20).

With the printed antenna PVC foils can be usually printed starting from a thickness of 100 µm. The antenna structure is printed with a silver paste using the silk-screen printing technique and dried afterwards. Sometimes the printing cycle is run twice in order to increase the layer thickness from approximately 10 to 20 µm. In contrast to the etched antenna here it is problematic to implement very fine conductive structures. Here conductive lines widths of at least 150 µm are usual. The printed antenna, contrary to the other antennas, can be manufactured under very economical conditions since the printing process is available in the smart card manufacturing process. A restriction is the resistance value of the antenna. For a simply printed coil this is from 30 to 50 Ω. The other types of antennas have a resistance value of approximately 5 Ω, which is necessary for some contactless systems. Naturally the resistance value can be degraded by increase of the layer thickness of the printed antenna. A disadvantage thus is that production and material costs are increased with the necessary multiple printing to reduce the resistance values down to the values of the etched or wired antenna.

For a future production method a system is under development which makes the etched antenna more environmental and more economical. Applying a full surface copper foil is to be allotted by selective separating of the copper to the layout of the antenna. So the production price for the etched antenna could be reduced. Using this procedure only very thin copper can be separated, which again increases the resistance value of the antenna and leads to a loss of range of the contactless smart card. Besides copper, aluminium could be used as a cheap material for the antenna.

For different antenna materials there are different possibilities for connecting the contactless chip module with the antenna.

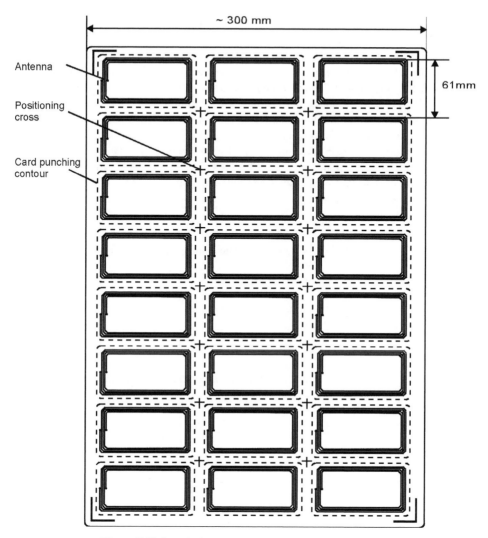

Figure 7.20 Sample for an antenna sheet with a 3 * 8 antenna

7.2.2   Connecting technologies for contactless smart cards

The different types of antennas described in Section 7.2.1 need partially different connection techniques.

In the following the welding connections between the contactless chip module and etched or wire embedded antennas are described. For the printed antenna there is a special connection technique — the so-called cut crimp technology.

Section 8.3 shows the flip-chip connection technology based on the contactless smart card.

First we explain the connection technique between the wire antenna, which was produced in the embedding technique, and the contactless chip module. Using this connection technique the contactless chip module is connected with the wire antenna by means of micro welding. Figure 7.21 shows the structure in principle of a contactless smart card manufactured using this connection technique.

As the first production step an intermediate foil **1** with the same thickness as the carrier tape of the contactless chip module is applied on the carrier foil. The intermediate foil **1** is punched out in the area of the contactless chip module. The punching hole has the size of the chip module contact plate. Next, one drop of adhesive is spread exactly in the position where the contactless chip module is to be later adjusted on the carrier foil. This is necessary in order that with the following production steps the contactless chip module is not shifted from its position.

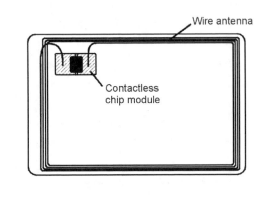

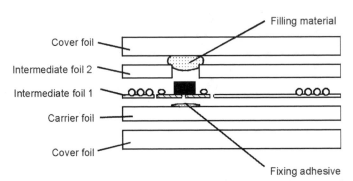

Figure 7.21 Structure in principle of a contactless card and the special binding of a contactless chip module by means of micro-welding to the embedding-technique manufactured antenna

Subsequently the contactless chip module is punched out from the carrier tape and placed on the adhesive drop. Afterwards a rubber stamp presses the chip module lightly into the adhesive, so that the chip module is fixed.

The next production step is connecting the isolated copper antenna wire ($\varnothing$ 125 μm) with the contacts of the contactless chip module. Using a special tool the isolated antenna wire in the connection area of the contactless chip module is stripped and connected permanently with the contactless chip module through micro-welding by means of ultrasonics. Subsequently, the antenna wire in the geometry of the later antenna in the embedding technique (for this see Section 7.2.1) is applied on the intermediate foil **1**. If the antenna is completely placed, the end of the antenna wire is welded to the second antenna contact of the contactless chip module.

Now another intermediate foil **2** is applied, which is left blank in the area of encapsulation of the contactless chip module. The thickness of this foil depends on the height of the encapsulation of the contactless chip module.

Figure 7.22 Placement robot for applying a contactless chip module on an etched antenna (source: Mühlbauer)

Depending upon the individual procedures of the smart card manufacturer, within the area of the encapsulation the gap between the punched-out recesses of the intermediate foils **1** and **2** and the contactless chip module is filled up with a filling compound. This prevents the formation of unevenness on the surface of the cover foil in that particular area.

Now at least one cover foil is added at the top and at the bottom in order that the thickness of 840 µm maximum required by the ISO is achieved. The majority of cover foils have printed artwork. Subsequently, the whole foil package is laminated in order to permanently connect the individual foils. Afterwards the individual contactless smart cards from the laminated sheets are punched out and output-controlled electrically and optically. Finally, the smart cards are ready for dispatch or personalization.

A further possible connection technique is the soldering technique between the contactless chip module and an etched antenna.

For this, as shown in Figure 7.22, a roll with etched antennas is used instead of the wired antenna.

Free punching is in the area between the two electrical contacts of the antenna into which the contactless chip module will be placed later (see Figure 7.23 and Figure 7.24). On top of the antenna windings and around the antenna connections an insulating covering is applied via silk-screen printing in order that the contactless chip module does not cause a short circuit after insertion into the free punched area.

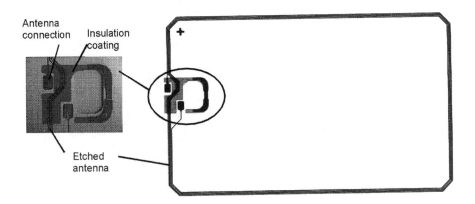

Figure 7.23 Example of an etched antenna with window enlargement of the connection areas and the isolation of the antenna windings

Next a soldering paste is spread with an automatic dispenser on the two antenna connections. Subsequently, the contactless chip module is punched out from the chip module carrier tape as shown in Figure 7.24 and Figure 7.25, and inserted into the free punching with the encapsulation pointing downward.

The free punching in the antenna foil also serves as adjustment for the contactless chip module. Subsequently, with two soldering points over the contacts of the contactless chip module heat is brought into the connection and the soldering paste melts. During cooling a permanent connection between the contacts of the contactless chip module and antenna contacts develops. Then the electrical functionality of the antenna with the chip module is checked and the equipped foil is rolled up or cut and destacked into single sheets of 3 * 8 antennas.

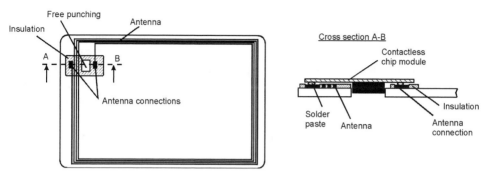

Figure 7.24 Schematic representation of a soldered connection between the contactless chip module and etched antenna

The assembly system described can equip three antennas being situated next to each other at the same time.

Now the equipped antennas can, as already shown in Figure 7.21, be laminated to a finished contactless smart card with different plastic foils.

A further variation technique is the connection of a contactless chip module using the 'cut crimp technology' of the Multitape company.

Figure 7.25 Assembly of an etched antenna with a contactless chip module (source: Mühlbauer)

Using the cut crimp technology printed and etched antennas can be contacted. The only restriction of this procedure is that the contactless chip module must have lead-frame contacts as carrier tape. Contactless chip modules with FR4 or another material as carrier tape cannot be used for this connection technique. Figure 7.26 schematically shows the cut crimp technique. The process steps with this connection technique are as follows:

First a pointed thorn picks into the two metal contacts of the contactless chip module and two crowns are formed. Subsequently, the contactless chip module from the chip module carrier tape is punched out and pressed from the reverse side into the PVC foil and then through the antenna connections. The points of the crowns penetrate into the plastic foil and the above-lying antenna contacts. Next the out-standing crown points are bent with a flat stamp in such a way that the crowns create a permanent mechanical connection with the antenna contacts.

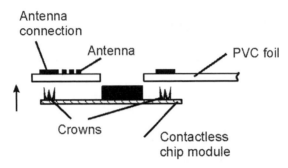

Figure 7.26 Schematic representation of the cut crimp technology

An advantage of this procedure consists of the fact that the isolation over the antenna is omitted because the contactless chip module is underneath the antenna and the substrate serves the antenna (PVC) as insulator (see Figure 7.26). Additionally, the electrical connecting is done without dispensing any conductive pastes.

7.3   Dual-interface smart cards

The dual-interface smart card (also see Sections 1.3.2 and 6) is a new, interesting product which is being introduced into the market with many pilot projects worldwide.

Dual-interface smart cards have the advantages of the contact smart card with the convenience of the contactless smart card.

From the point of view of the production technique, the dual-interface smart card is particularly interesting. Here the mounting technique, laminating technique, antenna technique and connection technique have to compete among themselves. In the following, combinations of connection techniques are described, e.g. embedded antenna with laser soldering, electric conductive paste with printed antenna and anisotropic heat-activatable adhesive with etched antenna. In addition, a manufacturing method is described which is based on the laminating technique.

7.3.1   Exposing of the antenna contacts in the card body

Before exposing of the antenna contacts in the card body is described, a few words about the antenna itself are called for.

As already mentioned in Section 7.2.1, there are restrictions in the position, geometry and size of the antenna on a smart card.

For dual-interface smart cards the main restricted fort is the area which includes the embossing. In this area the antenna could become damaged during the embossing procedure (also see Figure 7.18). Additionally the antenna must correspond to the antenna specifications published by the chip manufacturer such as antenna resistance, inductance and quality of the antenna. Regarding the position of the antenna contacts, they must be placed under the contacts of the chip module. The position of the chip module is described in ISO 7816, which clearly reduces the free selection of position for the antenna contacts.

Most connection techniques for the production of a dual-interface smart card require the antenna connections in the card body to be opened by milling. This is necessary for connecting the antenna contacts present on the back side of the chip module, electrically with the antenna contacts in the card body.

However, the antenna contacts in the card body cannot always be expected at the same depth (z-direction) from the card surface because of the thickness tolerances of the antenna and the plastic foils used.

It is helpful if the milling station which opens the antenna contacts has automatic antenna recognition. A possibility for managing this involves regulating the feed of the milling spindle in the z-direction in such a way that the downward movement is stopped when achieving the antenna contacts.

A technical solution is that the base plate of the milling station contains an antenna which induces a field. As soon as a card body arrives at the milling station, the resonant frequency is measured with a spectrum analyser (see Figure 7.27).

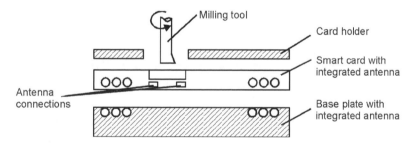

Figure 7.27 Automatic antenna contact detection with automatic depth regulation of the milling tool

Subsequently, the milling tool mills the plastic down to a pre-defined depth, which is a little bit above the antenna contacts. After this the milling tool works deeper, step by step in small steps, whereby the frequency spectrum analyser measures the resonant frequency constantly. As long as no frequency distortion is measured, the milling tool steps in the z-direction in small distances. This loop, measuring and milling, occurs until the resonant frequency changes. This is an indication that the antenna contacts are now uncovered. The procedure is very time-intensive, therefore one tries to manage without automatic antenna recognition by increasing the milling accuracy of the milling machine and increasing the layer thickness of the antenna contacts.

Now the two antenna contacts (see Figure 7.28) are uncovered and ready for bonding of a dual-interface chip module.

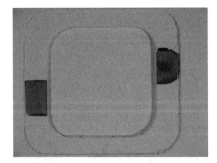

Figure 7.28 Milled cavity for a dual-interface chip module with antenna connections

7.3.2 Connection technology for the antenna

After the antenna contacts in the card body are opened, the electrical connection of the antenna contacts to the dual-interface chip module can be set up.

As a first example of the connection between the antenna and chip module we describe the connection technique involving laser soldering. Here a card body with an internal wired antenna is used. The location of the antenna indicates a unique feature within the area of the later junction point. The antenna wire is not led to the junction point, but shifted in a wavy line (see Figure 7.29). This unique approach of the antenna wire should protect the later electrical connection against mechanical tensile loads and increase the available surface for electrical connection.

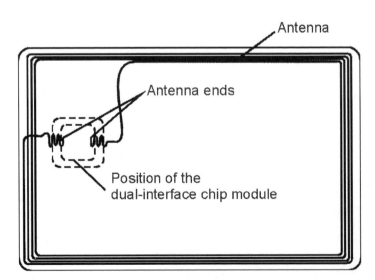

Figure 7.29 Position of one wiring-technique-produced antenna for a dual-interface smart card

In a second production step, a silver solder paste is spread on the two opened antenna contacts and the dual-interface chip module is inserted into the cavity. The connection of the chip module and card body can be made with heat-activatable adhesive or a liquid adhesive system.

In the case of usage of a heat-activatable adhesive the area of the chip module antenna contacts naturally has to be left open, otherwise the adhesive functions as an insulator between the chip module contact and antenna contact.

When the chip module is fixed in the cavity, the silver solder paste is warmed up through the substrate (FR4) using a short laser impulse and thus the connection between the antenna contacts of the chip module and the antenna is established (see Figure 7.30).

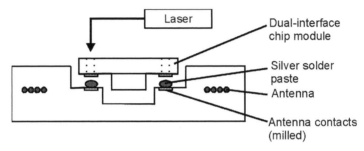

Figure 7.30 Principle sketch for the connection technique of a dual-interface chip module with an antenna inside a card body through laser energy

In order that the laser beam can penetrate through the dual-interface chip module to the junction point, this dual-interface chip module must exhibit a unique characteristic in relation to other dual-interface chip modules. A hole must be present in the contact plate above the two junction points to the antenna in order that the laser light can pass the chip module and warm up the silver solder paste (see Figure 7.31). The diameter of the hole in the contact plate must correspond to the wavelength of the laser light used.

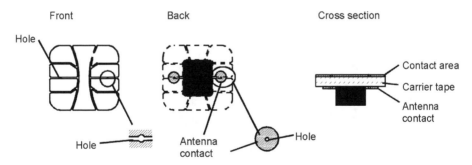

Figure 7.31 Structure of a special dual-interface module for the connection technique by means of laser soldering

This connection technique is a quick option for making a permanent electrical connection. The only problem consists in positioning the laser exactly over the hole in the chip module. Also the laser energy intensity has to be selected in order that it does not cause thermal deformation on the smart card rear side. The dual-interface chip module itself can be fixed in the cavity through a heat-activatable adhesive or a reactive adhesive.

A further possibility for manufacturing a dual-interface card consists of using silver conductive adhesives as connecting media between the dual-interface chip module and silk-screen printed antenna. The printed antenna contacts must be directly below the dual-interface chip module contacts and on one level (variant **I** in Figure 7.32). The second version would be to through-supply the antenna underneath the glob top of the dual-interface chip module (variant **II** in Figure 7.32). The antenna contacts would also have to be at this level.

In order to implement variant **I**, the antenna must be distributed on two levels in the structure of the card body (see Figure 7.33). This is necessary as the antenna ports **D** and **D'** are on one side of the antenna. In order to connect the antenna port **D** with the remaining antenna, a junction line would have to be led across the antenna lines. This would lead to a short circuit. This problem can be overcome if the antenna is distributed on two foil levels.

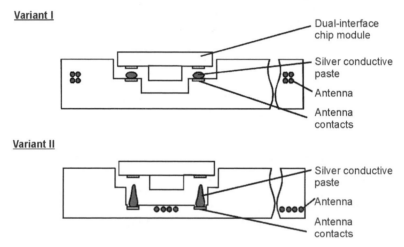

Figure 7.32 Two versions for the position of the printed antenna in the card body for the production of the electrical connection of the antenna contacts with the dual-interface chip module

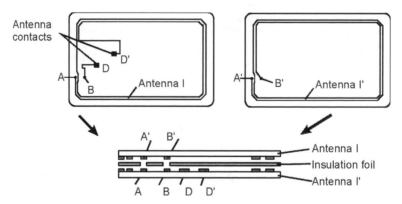

Figure 7.33 Printed antenna on two levels in the structure of the card body distributed for the bypass of the antenna contacts

An antenna, which consists for example of four turns, is distributed in two turns each on two foils (see Figure 7.33). To enable the later electrical connection, the antenna print must be arranged face to face. Therefore between the antenna foil **I** and **I'** a thin insulation foil is positioned for electrical isolation. In order that the two section antennas can be connected electrically, a hole is punched into the insulation foil within the area of the ports **A** and **B**. Ports **A** with **A'** and **B** with **B'** are connected using the lamination process.

As the next production step the antenna ports **D** and **D'** are free-milled. Afterwards, using a dispenser a silver conductive paste is spread on the antenna contacts. When pulling the dispensing needle away attention must be paid to the fact that the silver conductive paste pulls no threads leading to short circuits. In addition, the silver conductive paste should not reach the smart card surface, otherwise it may get dirty and stick with other smart cards when later destacking in the supply magazines.

The dispensed silver conductive paste should be checked with (for example, optically) a measuring system — the diameter and the position should be confirmed. This is particularly necessary with manufacturing variant **II** as without enough silver conductive paste the electrical contact does not take place. Subsequently, the dual-interface chip module is fixed in the card body using liquid adhesive systems or with a heat-activatable adhesive. When using heat-activatable adhesives the area where the antenna contacts on the dual-interface chip module has to be left blank, otherwise the silver conductive paste cannot establish an electrical contact between the two contacts (chip module/antenna).

When the dual-interface chip module is positioned in the cavity of the card body, the silver conductive paste is activated by being pressed down with a heated welding stamp. Depending upon the silver conductive paste used the finished card must harden for some hours at ambient temperature or in a heating oven, until the connection obtains its final stability.

As a last possibility for manufacturing a dual-interface card the use of a heat-activatable adhesive with anisotropic electrical characteristics is presented here (Figure 7.34).

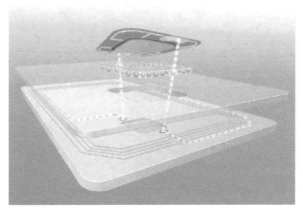

Figure 7.34 Principle of a dual-interface smart card with anisotropic electrical heat-activatable adhesive HAF8412 (source: TESA)

The principle of anisotropic leading systems consists of the fact that the electric current flows only in the z-direction — as can be seen in Figure 7.35.

In order that no electrical current flow can take place in the x- or y-direction, the distance d between the electrically conductive balls must be sufficiently large in order to prevent them from touching themselves mutually. In order establish the electrical connection between the dual-interface chip module and the antenna contacts, for example of the etched antenna, the heat-activatable adhesive is filled with silver-coated balls (see Figure 7.36). The balls used can be flexible (made from plastic) or rigid (made from glass).

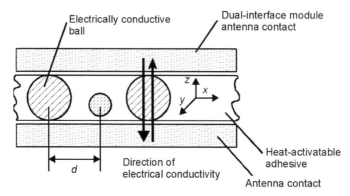

Figure 7.35 Principle of an anisotropic conductive heat-activatable adhesive

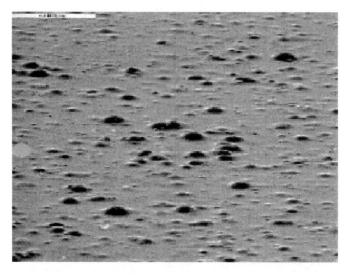

Figure 7.36 Electron micrograph picture of the surface of an anisotropic conductive heat-activatable adhesive (source: TESA)

A problem exists with the already described etched antenna in that the two antenna contacts for the dual-interface chip module are on the inside of the antenna windings. This problem is by partially applying solved an insulating varnish between contacts **A** and **B** (see

Figure 7.37) on the antenna windings. Over the isolated area electrical connection between contacts **A** and **B** is established through a silk-screen printed silver conductive paste.

The only restriction with this type of connection technique is that the antenna contacts must be at a certain level (*z*-depth) in the card body. For a dual-interface chip module with a carrier tape thickness of 200 μm the antenna contacts should be at a depth of 230 to 250 μm from the card surface. The exact value depends on the thickness of the anisotropic adhesive used.

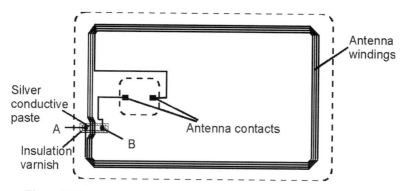

Figure 7.37 Example of an etched coil layout for a dual-interface smart card

The following production steps are the same as for the mounting technique described in Section 7.1. First the anisotropic adhesive is up-laminated on the plasma pre-treated dual-interface chip module carrier tape. Contrary to the procedure with silver conductive paste, it is not necessary to free punch the adhesive within the area of the contacts — this would prevent electrical conductivity in a place where it is needed; however, it has the advantage that nothing has to be modified in the usual machine operational sequence of the regular mounting technique.

After lamination of the adhesive, the dual-interface chip module is brought into the card body with the free-milled antenna contacts, and afterwards fixed in the cavity of the card body using a heated stamp. Subsequently, using a heated stamp the electrical and mechanical connections between the chip module and antenna contacts are carried out in one production step.

Now the dual-interface smart card is fully usable and there is no need for hardening afterwards.

The advantage of the connection technique using anisotropic heat-activatable adhesives in relation to the other two described procedures is in the fact that the available manufacturing machines of the mounting technique can be used with only slight modifications (no additional dispensing stations). Anisotropic heat-activatable adhesives are available, for example, from the TESA company.

As a final possibility for manufacturing a dual-interface smart card a production technique is described based on the laminating technique. Figure 7.38 shows the structure in principle of a dual-interface smart card manufactured using the laminating technique. Here, for example, on a foil of 3 * 24 an antenna window of the size of a chip module encapsulation has to be punched out between the antenna contacts.

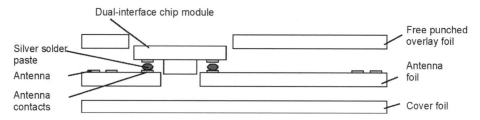

Figure 7.38 Construction of a dual-interface smart card produced from the lamination technique

The foil inclusive antenna must be thick enough in order that the encapsulation of the dual-interface chip module fits in the depth z. Naturally, the foil thickness can be divided into several foils but every foil has to be provided with a free punched window.

Next a solder paste is spread on the antenna contacts, afterwards the dual-interface chip module is punched out of the chip module carrier tape and inserted into the punched-out window of the foil.

Subsequently, by putting two soldering needles on the contact area of the chip module the silver paste between the antenna contacts and the chip module is melted and hence a mechanical and electrical connection to the antenna is established. Finally, a punched-out overlay foil is applied (see Figure 7.38). The punching out has to be somewhat larger than the external dimensions of the chip module contact plate. The larger the tolerances are when placing the chip module and the position accuracy of the recess the larger the circulating gap has to be. If the recess is arranged too close, the overlay foil may flow over the contacts with the later laminating process and could possibly lead to an impairment of the electrical functionality of the external contacts.

Finally, a foil is applied on the rear side. Both this foil and the overlay foils may be divided into several foils.

Next the mounted foil package is laminated.

If the temperature load for the chip module can be reduced, the individual foils can also be connected by using a cold adhesion method.

As the last production step the individual dual-interface smart cards must be punched out from the laminated foil. The punching out has to be very precise as the later positions of the electrical contacts of the chip module are referenced to the edge of the smart card. Now the punched-out dual-interface smart card only has to be checked electrically and optically (position of the ISO contacts).

This manufacturing method is more problematic than the manufacturing methods described previously because of the multiple tolerances which have to be considered. Additionally, many expensive dual-interface chip modules may be damaged in one go by possible errors within the laminating process.

We have provided an overview of some manufacturing methods for the production of a dual-interface smart card. Given that the dual-interface smart card still awaits its mass introduction on the market and insignificant numbers have been produced, the debate is still open as to which manufacturing method will establish itself and whether any further manufacturing methods will be developed.

8 Smart cards without chip modules

In addition to smart cards with chip modules there are some new packaging technologies like CSP (chip size package) without chip modules. These new packages also use new connecting technologies for smart cards like flip-chip technology. This movement is especially seen for low cost memory smart cards (telephone cards) and for contactless smart cards. In this chapter we consider some established solutions on the market.

8.1 MOSAIC

The MOSAIC (microchip on surface and in card) manufacturing method from the Solaic company was one of the first procedures for the production of smart cards with chips without chip modules.

The production process flow (see Figure 8.1) is as follows: The chip is brought directly from the sawed wafer into the smart card. For this the card body is locally heated up by a laser beam. Subsequently, the chip is pressed into the partially softened card body until the chip surface and the card surface are at the same level. After cooling of the smart card the chip is permanently fixed in the plastic material.

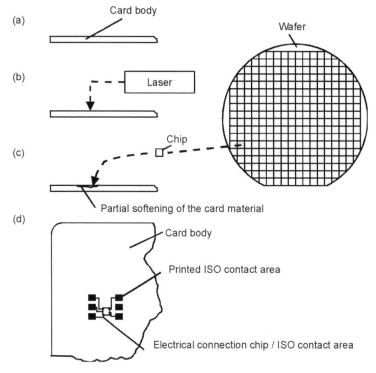

(a)   Card body

(b)   Laser   Wafer

(c)   Chip

Partial softening of the card material

(d)   Card body

Printed ISO contact area

Electrical connection chip / ISO contact area

Figure 8.1 Production flow of a MOSAIC smart card

As the next step, the ISO contact areas and the connection lines to the chip pads are printed with a silver conductive paste on the card surface. Figure 8.2 (left picture) shows the ISO contact areas enlarged. In the right picture the connections to the chip pads are shown enlarged.

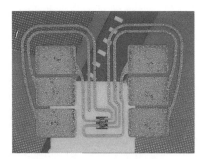
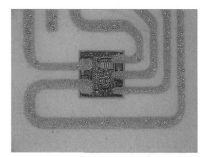

Figure 8.2 ISO contact areas (left) of a smart card and on the right the connection of the contact area to the chip pads

For the last manufacturing step, an insulating varnish is applied, which covers the access lines and the chip in such a way that they are protected against mechanical and chemical damage. Naturally, the ISO contact areas are not covered by the insulating varnish. An advantage of this technique is that fewer manufacturing steps from the wafer to the finished smart card are necessary than with the conventionally used chip modules. Additional materials such as FR4, copper and gold, which are used for the production of the chip modules are saved.

One disadvantage of the technique is that only chips with a total size of approximately 1 mm² can be inserted as the danger of breakage of the unpackaged chip increases with larger chip size. A further disadvantage is that precise printing machines are necessary for the bonding of the chip to the contact areas.

At the present moment the MOSAIC technique is used for telephone cards for German Telekom.

Next we look at a technology which additionally protects the chip and whereby the chip can be nevertheless directly processed from the wafer.

8.2 Shellpack

Shellpack is a packing form in which no chip module of the conventional form is used. Shellpack technology is offered by the Shellcase company.

Simply put with Shellpack technology the chip is embedded (better to say the whole wafer) between two glass discs. The chip contacts are lead outward to the glass surface. The contacts can have any form and position on the glass surface. The packing manufactured in such a way can be brought into the card, for example, for the production of a contactless smart card using flip-chip technology (see Section 8.3). Figure 8.3 shows the principle structure of a Shellpack package.

The typical thickness of a Shellpack package is approximately 350 µm.

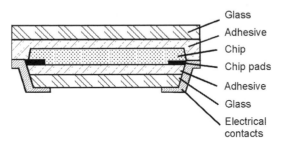

Figure 8.3 Cross section of a Shellpack package

In the following the most important process steps of how a Shellpack package is made from a wafer are described.

In the first step for Shellpack technology the electrical connection of the chip pads must be extended outward into the sawing street. This is achieved by lithographic procedures and metal deposition.

As the next manufacturing step, a flat special glass disc with an epoxy adhesive is glued on the active side of the chip. Afterwards the wafer becomes back-lapped from the rear side from typically 360 µm down to 100 µm.

Now a film of varnish is applied on the rear side of the wafer, which releases the sawing street, and afterwards the chips are separated from each other with chemical etching. The single chip does not lose its position as it is bonded to the glass disc.

Then epoxy adhesive is applied on the rear side of the wafer and a further glass disc is bonded on. Now a system is present in which the chips are sorted between two glass discs (see Figure 8.4).

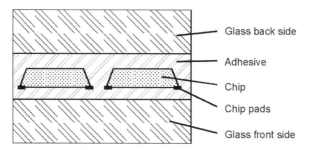

Figure 8.4 Embedding of the wafer between two glass discs after separation of the individual chips by chemical etching

In the next process step using lithographic and etching processes trenches are made into the glass disc which lies on the side of the electrical contacts. The trenches are deep enough to expose the electrical connections (see Figure 8.5).

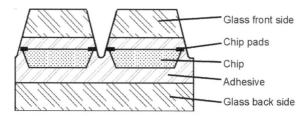

Figure 8.5 Release of the electrical connections in the sawing street

Subsequently, a layer of metal is applied on the upper glass disc, at the external walls of the ditches and on the electrical chip pads. Hence the electrical connection to the external glass disc is set. The metallization layer on the glass surface must only be arranged after the non-standard design of the requested contacts (see Figure 8.6).

For later contacting on the metal surface a thin nickel layer is deposited.

Finally, the individual chips are separated by the sawing process and are now prepared for further use in the smart card or in a chip module.

The machines used for the Shellpack procedure are well known from normal semiconductor technology. At present the Shellpack procedure is not used in large volume for the smart card.

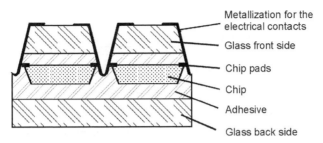

Figure 8.6 Extension of the chip pads on the top of the glass disc

Next we present a connection technique which can be used, for example, for contacting of the packing manufactured in Shellpack technology with which, in addition, unpacked chips are also processable.

8.3 Flip-chip connection technology

Here we describe a connecting technology with which a smart card without a chip module can be manufactured using so-called 'flip-chip technology'.

Flip-chip technology was developed in order to compensate for the disadvantages of conventional wire bonding like, for example, large space requirement by the wire-rod guide on the substrate.

One of the first flip-chip procedures well known as 'solder-bumped flip-chip technology' or also as C4-Prozess (controlled collapse chip connection) was introduced in 1960 by IBM.

The C4-Prozess requires complex pre-treatments of the chip contact areas, and process temperatures up to 320 °C are achieved. For application in PCs (personal computers), however, this connection technique was an enormous step towards miniaturization, high component density and cost savings.

For application in a smart card, in particular for the contactless smart card, the high process temperatures of the C4-Prozesses are not applicable. PVC, for example, already softens at approximately 60 °C; in addition, the connection technique applied with the C4-Prozess is too cost-intensive for a price-driven product like the contactless smart card.

In the meantime many different procedures were developed which are economical and primarily used for the contactless smart card.

Before going into the detail of the different connecting technologies and methods for pre-treatment of the chip pads let us have a quick look at what flip-chip (FC) technology is:

Figure 8.7 shows the principle of FC technology. The main difference between wire bond technology and FC technology is that the side of the chip where the pads and electrical circuits are looks towards the substrate, not away as in conventional wire bond technology.

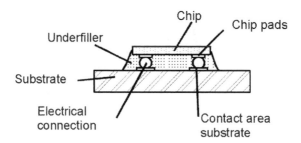

Figure 8.7 Principle of flip-chip (FC) connecting technology

The flip-chip structure is basically quite simple. Between the pads of the chip and those of the substrate is an electrically conductive material which ensures the electrical connection.

By use of a liquid adhesive, the so-called 'underfiller', the system chip/substrate and the electrical connection is mechanically held together.

The footprint of a flip-chip 'package' is only of the size of the chip itself.

To establish the electrical connection between the chip pads and the substrate different solutions are possible (see Figure 8.8).

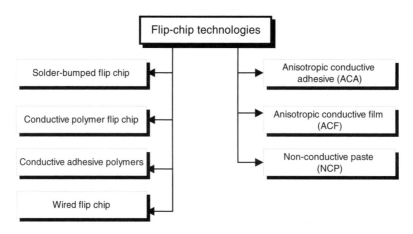

Figure 8.8 Overview of some flip-chip connecting technologies

First the chip pads have to be coated with an intermediate layer for the later electrical connection.

This intermediate layer serves among other things to protect the chip pads which consist of aluminium against humidity. Additionally the intermediate layer is more suitable as connecting material for later contacting than aluminium.

The necessary process step for it is undertaken at the wafer level. First, the aluminium pads are chemically cleaned and submitted to a zinc pre-treatment. Subsequently, the wafer is dipped into a nickel bath and an approximately 10-µm-thick nickel layer separated on the pad surface. For improvement of the later contact quality an additional thin gold layer is applied. The so-called under bump metallization (UBM), developed in such a way, is shown schematically in Figure 8.9. Figure 8.10 is a magnification of UBM on a test structure.

This procedure is offered, for example, by the PacTech company.

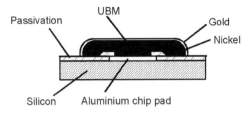

Figure 8.9 Under bump metallization (UBM) of a chip pad with nickel and gold

 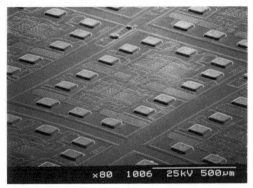

Figure 8.10 Under bump metallization (UBM) of a test structure (source: PacTech)

The UBM can reach a height up to 20 µm. In addition to the nickel/gold UBM there is also the possibility of using palladium coverage for the chip pads. This technology is provided by the KSW company.

Now the wafer is prepared for usage in flip-chip mounting technologies.

Using two examples we show how flip-chip mounting technology can be used to make contactless smart cards.

First, in silk-screen printing an electrical conductive silver polymer paste is applied on the Ni/Au UBM and hardened afterwards. This production step is still undertaken at wafer level.

The use of polymer pastes has the advantage in contrast to the usual fluxes that later process steps can be operated with substantially lower processing temperatures and process pressure.

Up to this production step the entire production is undertaken at wafer level. Through this very high numbers of items can be pre-treated economically.

As the next step, the wafer is applied to an adhesive foil and sawed afterwards. Now the individual chips on the adhesive foil (also see Section 5.1) are individually available. Afterwards the antenna contacts on the PVC foil are prepared for applying the chips. For this a polymer conductive paste is spread at the two antenna contacts. The chip is applied towards the antenna contacts with the electrical links to the antenna. When placing the chip it is slightly warmed up, hence the connection between antenna and chip is pre-cured. The advantages of polymer conductive pastes are that the activation temperatures are below 60 °C and the PVC foil is not thermally deformed.

However, since this connection is mechanically very unstable, a liquid adhesive system (so-called underfiller) is spread around the chip. The underfiller due to the capillary effect flows into the gap between the chip and PVC foil. In the next process step the underfiller is cured.

Afterwards the finished equipped PVC foil can be stacked up by additionally applied plastic foils to the required card thickness and can be finished with the conventional laminating procedure.

A further possibility, in order to make an electrical connection using flip-chip technology, is the usage of anisotropic conductive adhesives (ACAs). ACAs are adhesives which are filled with small conductive particles, for example, gold-coated plastic balls with a diameter of approximately 5 µm. The advantage of ACA involves the fact that the electrical conductivity is only possible in one direction: in the Z-direction (also see Section 7.3.2), but not in the X- or Y-direction (see Figure 8.11).

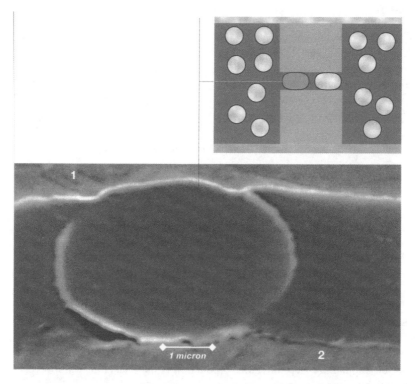

Figure 8.11 Enlargement of an ACP (anisotropic conductive paste) connection (source: DELO)

Also with this type of connecting the chip pads should receive a pre-treatment (UBM) of the aluminium surface. Staying with the example of the contactless card, afterwards a drop of ACA is dispensed on the antenna contacts as shown in Figure 8.12. Subsequently, the chip is placed onto the antenna contacts and pressed against them under a small force. At the same time the anisotropic adhesive is activated and hardened from the foil rear side using a UV lamp. In order that curing can take place through the PVC foil the foil must be UV light transparent, which leads to a restriction with the foil selection. If other ACA material is used, then the curing can be undertaken by temperature or microwaves.

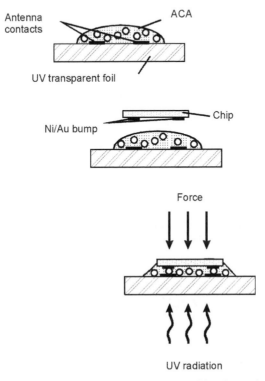

Figure 8.12 Flip chip contacting for the contactless smart card with anisotropic conductive adhesives (ACAs)

The hardened ACA at the same time serves as mechanical stabilization between the chip and plastic foil.

Finally, some additional foils have to be added to the antenna foil to reach the final card thickness.

The connection technique with ACAs which can be activated with UV radiation offers a reduction of process steps in relation to conventional connecting techniques, and will be a cost-efficient solution for the production of contactless smart cards.

In the preceding chapters the most diverse versions of card bodies and chip modules were presented. However, in order that new smart cards are suitable for the high requirements, extensive reliability tests must be executed. The most usual methods for that are presented in the next chapter.

9 Reliability tests for chip modules and smart cards

In daily use the smart card and in particular the chip module are exposed to a variety of mechanical, physical and chemical loads.

For example, smart cards are kept frequently in a purse which is then put into the hip pocket. The smart card is constantly bent by the movements, and the edges of coins can press on the chip module.

Or with application of the smart card in a mobile telephone: If the mobile telephone is situated in a car, which stands in the sun and is strongly heated, temperatures up to 80 °C can be reached.

In order that a smart card can fulfil all the different requests, national (DIN) and international standards (ISO, MIL-STD) were introduced where the testing methods and test criteria are described.

The range of reliability tests performed depends on different facts. The first and most important factor is naturally customer requirements. For example, for smart cards which are used in mobile phones for GSM (Global System for Mobile Communication) the GSM specification applications are used.

A further factor is the internal requirements of the semiconductor and smart card manufacturers as well as the smart card-relevant national and international standards.

In the following Sections, we introduce some of the most important electrical and physical reliability tests.

First, however, we give some general explanations for quality assurance.

9.1 Quality control, AQL, LTPD

As shown in previous chapters, a smart card consists of many individual components, e.g. plastic foils, magnetic stripes, holograms, printing inks and of course the chip module.

Usually all these individual components undergo an inspection for incoming goods when delivered to the smart card manufacturer. In the following, we explain some fundamental ideas and procedures for inspection of incoming goods on the basis of chip modules.

For example, for chip modules supplied to the smart card manufacturer, usually 100 % electrical functional control of each individual chip module is not undertaken as the chip modules are usually already output-controlled by the chip manufacturer to 100 % on wafer level and to 100 % electrically in the chip module itself. Additionally, a function test would lead to very long testing times for microcontroller chip modules with large EEPROM memory and would cause unnecessary costs. The electrical test can be undertaken substantially faster with the chip manufacturer than with the smart card manufacturer as the chip is with the smart card or module manufacturer in the so-called 'user mode'. The chip is switched into the 'user mode' by the manufacturer for safety reasons as otherwise the software in the chip would be accessible easily by others. In this mode, fast test routines are not available as is the case before switching into the 'user mode'.

Due to time and cost reasons all tests cannot be executed; however, a certain level of assurance is desirable with regard to the quality of the delivered chips or chip modules hence the chip modules are tested in a random sampling way according to the specified demand.

The acceptable quality limitation level (AQL) is defined as the maximum proportion of defect units in per cent. The AQL value is determined between suppliers and customers. Afterwards, a certain number of units are tested from the inspection lot (for example, of a transport reel with chip modules). Using the single sampling plan the error number which can be accepted can be finally found.

In order to clarify the procedure, the example described beginning with the delivered chip modules is next explained in further detail.

The starting quantity for the smart card manufacturer is a transport reel with, e.g. 15,303 functional chip modules.

The code letter M, at a normal inspection level II, can be taken from Table 9.1 with a lot size of 15,303 units. Next the required sample size has to be determined. From Table 9.2, a sample size of 315 chip modules can be found with the code letter M. The 315 chip modules must be tested according to the coincidence principle distributed from the whole transport reel.

Table 9.1 Sample size code letters

Lot size		Normal inspection level		
		I	II	III
1-	8	A	A	B
9-	15	A	B	C
16-	25	B	C	D
26-	50	C	D	E
51-	90	C	E	F
91-	150	D	F	G
151-	280	E	G	H
281-	500	F	H	J
501-	1200	G	J	K
1201-	3200	H	K	L
3201-	10,000	J	L	M
10,001-	35,000	K	M	N
35,001-	150,000	L	N	P
150,001-	500,000	M	P	Q
500,001>		N	Q	R

Now the acceptance number c and the rejection number d have to be determined from Table 9.2. In addition the AQL value which was determined between chip manufacturers and smart card manufacturers is needed. For example, an AQL value of 0.65 is used here. In the case of a code letter M and an AQL value of 0.65 an acceptance number c of 3 and a rejection number d of 4 results. This means that out of the delivered goods parts a maximum 3 chip modules out of the 315 tested chip modules may not correspond with the quality required. However, if four or more chip modules are found, which do not correspond to the specified delivery quality, the transport reel can be returned to the manufacturer or an intensified test has to be submitted.

The specified delivery quality may contain the optical, geometrical and electrical characteristics of the chip module.

Higher AQL values for the number of testing cards can be seen in Table 9.3

Table 9.2 Single sampling plan (master table part I)

Sample size code letter	Sample size	Acceptable quality level (AQL) (serve inspection)													
		0.010	0.015	0.025	0.040	0.065	0.10	0.15	0.25	0.4	0.65	1.0	1.5	2.5	4.0
		c d	c d	c d	c d	c d	c d	c d	c d	c d	c d	c d	c d	c d	c d
A	2														
B	3														
C	5														
D	8														▼
E	13													▼	0 1
F	20												▼	0 1	1 2
G	32											▼	0 1	1 2	2 3
H	50										▼	0 1	1 2	2 3	3 4
J	80									▼	0 1	1 2	2 3	3 4	5 6
K	125								▼	0 1	1 2	2 3	3 4	5 6	8 9
L	200							▼	0 1	1 2	2 3	3 4	5 6	8 9	12 13
M	315						▼	0 1	1 2	2 3	3 4	5 6	8 9	12 13	18 19
N	500					▼	0 1	1 2	2 3	3 4	5 6	8 9	12 13	18 19	▲
P	800				▼	0 1	1 2	2 3	3 4	5 6	8 9	12 13	18 19	▲	
Q	1250			▼	0 1	1 2	2 3	3 4	5 6	8 9	12 13	18 19	▲		
R	2000		▼	0 1	1 2	2 3	3 4	5 6	8 9	12 13	18 19	▲			
S	3150	▼	0 1	1 2	2 3	3 4	5 6	8 9	12 13	18 19	▲				

▼ Use first sampling plan below arrow. If sample size equals or exceeds lot size do 100 % inspection

▲ Use first sampling plan above arrow c = acceptance number; d = rejection number

Contrary to the AQL value, which describes the risk for the manufacturer that a part of the supplied product is returned, the LTPD value (lot tolerance per cent defective) describes the risk which the customer takes.

Frequently, the LTPD value is used as a failure criterion for the reliability tests. Table 9.4 shows an extract from the LTPD sampling table.

If, for example, a LTPD value of 7, also called the confidence level, is selected from the first row out of Table 9.4, a sample size of 75 is selected from the column below the 7. That leads to the fact that only two units are allowed to fail a reliability test. The value of 2 is found in the first column of the LTPD sampling table. If three units fail the reliability test undertaken, then the whole test sample is failed.

Table 9.3 Single sampling plan (master table part II)

Sample size code	Sample size	Acceptable quality level (AQL) (serve inspection) 6,5	10	15	25	40	65	100	150	250	400	650	1000
		c d	c d	c d	c d	c d	c d	c d	c d	c d	c d	c d	c d
A	2	▼			▼	1 2	2 3	3 4	5 6	8 9	12 13	18 19	27 28
B	3	0 1	▼	▼	1 2	2 3	3 4	5 6	8 9	12 13	18 19	27 28	41 42
C	5			1 2	2 3	3 4	5 6	8 9	12 13	18 19	27 28	41 42	▲
D	8	▼	1 2	2 3	3 4	5 6	8 9	12 13	18 19	27 28	41 42		
E	13	1 2	2 3	3 4	5 6	8 9	12 13	18 19	27 28	41 42		▲	
F	20	2 3	3 4	5 6	8 9	12 13	18 19				▲		
G	32	3 4	5 6	8 9	12 13	18 19		▲	▲	▲			
H	50	5 6	8 9	12 13	18 19	▲	▲						
J	80	8 9	12 13	18 19	▲	▲							
K	125	12 13	18 19	▲	▲								
L	200	18 19	▲	▲									
M	315	▲	▲										
N	500												
P	800												
Q	1250												
R	2000												
S	3150												

▼ Use first sampling plan below arrow. If sample size equals or exceeds lot size do 100 % inspection

▲ Use first sampling plan above arrow c = acceptance number; d = rejection number

The confidence level also enables a weighting of the individual tests. Low LTPD values lead to a higher demand for the quality of the product.

Today it is usual to pursue the 'zero-error strategy'. To the appropriate LTPD values the number of units under test is selected in such a way that failures are permitted to zero. For example, at a LTPD value of 20, eleven units have to be tested and with the tests undertaken no failures are allowed.

To show how differently reliability tests can be weighted and the acceptance boundaries for reliability tests, in the following sections some reliability tests for chip modules and smart cards are demonstrated.

Table 9.4 Extract out of the LTPD (Lot tolerance per cent defective) sampling table (source: MIL-S-1950)

LTPD (λ)	30	20	15	10	7	5	3	2	1.5	1
0	8	11	15	22	32	45	76	116	153	231
1	13	18	25	38	55	77	129	195	258	390
2	18	25	34	52	75	105	176	266	354	533
3	22	32	43	65	94	132	221	333	444	668
4	27	38	52	78	113	158	265	398	531	798
5	31	45	60	91	131	184	308	462	617	927
6	35	51	68	104	149	209	349	528	700	1054
7	39	57	77	116	166	234	390	589	783	1178
8	43	63	85	128	184	258	431	648	864	1300
9	47	69	93	140	201	282	471	709	945	1421
10	51	75	100	152	218	306	511	770	1025	1541
11	54	83	111	166	238	332	555	832	1109	1664
12	59	89	119	178	254	356	594	890	1187	1781
13	63	95	126	190	271	379	632	948	1264	1896
14	67	101	134	201	288	403	672	1007	1343	2015
15	71	107	142	213	305	426	711	1066	1422	2133
16	74	112	150	225	321	450	750	1124	1499	2294
17	79	118	158	236	338	473	788	1182	1576	2364
18	83	124	165	248	354	496	826	1239	1652	2478
19	86	130	173	259	370	518	864	1296	1728	2591
20	90	135	180	271	386	541	902	1353	1803	2705
25	109	163	217	326	466	652	1086	1629	2173	3259

9.2 Electrical reliability tests

The smart card is a highly innovative product. The security functions are improved constantly; operation speed and memory capacity is increased; but at the same time the total chip area is reduced by decrease of the structure size.

In order to maintain improved product properties at a high quality level, extensive reliability tests are undertaken before a new product arrives on the smart card market.

In the following, the reliability tests are divided into physical — also so-called mechanical — and electrical tests.

With the electrical reliability tests described in the following text the chip and the chip module are tested for their functionality and mechanical and electrical stability in relation to the environmental conditions to which they will be exposed in daily use.

Reliability tests are executed often. There are tests which are undertaken on each chip or chip module produced and take place in a random sampling way (see Section 9.1). Tests are also available which are executed uniquely for product introduction.

Wafer outgoing control is a reliability test undertaken by the chip manufacturers. At the outgoing control every individual chip on the wafer is loaded with higher values (like voltage and temperature) as in normal usage of a chip.

With such wafer outgoing tests, initial failures as shown in Figure 9.1, are sorted out from the wafer. The chips used in the later smart card should be in the region of random failures. In this region, the user of the smart card can be sure that the smart card operates very reliably over a long period. Naturally, as with all consumer goods, there is a wear out failure period where increasing loss of functionality occurs. However, this period is far behind the normal usage time of a smart card.

The graph shown in Figure 9.1 plots the failure rate in relation to the lifetime of a smart card. Owing to the shape the curve is also called a bath-tub curve.

In addition to the repetitive tests, the chip also has to pass through a range of reliability tests for the chip module.

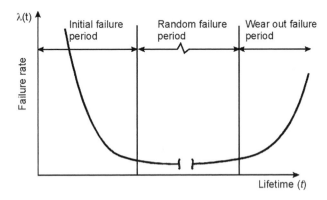

Figure 9.1 Failure rate curve for chips over a lifetime (bath-tub curve)

Table 9.5 lists a selection of electrical reliability tests for chips and chip modules. The standard belonging to a particular test is given in the table, which describes the test, and the last column indicates the usual test parameters as well as the duration.

Before we go into more detail regarding some reliability tests, some terms have to be defined:

'Lifetime' of a smart card means the time period after handing out the smart card to the consumer up to their application end.

Under 'actual working time' is defined as the time period in which the smart card, and in particular the chip, operates actively in the smart card reader.

The application conditions are described by the normal site conditions of temperature (25 °C) and relative humidity (50 %).

Here we describe some reliability tests in more detail from Table 9.5.

During the temperature humidity bias (THB) test the chip module is brought into a test chamber, which is operated with a test temperature of 40 °C and a relative air humidity of 93 %. The chip module remains for between 168 and 504 h in this environment. During the entire testing period, a supply voltage is applied at the chip module. The testing time depends on the later field of application of the smart card. With smart cards built for a short actual working time (for example, a pre-paid calling card) usually 168 h is enough. However, if the smart card is used e.g. as a cash card then the test is run for 504 h in order to reflect the longer actual working time.

Understandably, the question arises as to why the test of a cash card is carried out for only 504 h and not over 26,280 h (corresponding to permanent application over three years), which corresponds to the lifetime of a cash card. Such a long testing period would be neither practicable nor profitable. Therefore, so-called acceleration models were established in the semiconductor industry. With tightened up test conditions in relation to the usual operating conditions of everyday use, the test duration can be shortened substantially.

Table 9.5 Selected overview of electrical reliability tests

Reliability test	Method	Parameter/duration
Temperature humidity biased (THB)	IEC 68 parts 2 — 3	40 °C; 93 % rel. humidity; V_{cc} high; 168 — 500h.
Electrical resistance and impedance of contacts	ISO 10373	500 µA — 100 mA
Ultraviolet light	ISO 10373	15 Ws/cm²; 60 min each side
X-rays	ISO 10373	70 kV; 0.1 Gy
Magnetic field	ISO 10373	79,500 A/m
Life test	MIL-STD 883/1006	85 — 125 °C $V_{cc} = V_{ccmax}$; f_{clk} = Active
Electrostatic discharge (ESD)	MIL-STD 883/3015.7 or ISO 10373-3 ISO 10536-1, A3 ISO 10373-3	1.5 — 4 kV HBM; 100 pF; 0.5 kΩ 10 kV; 100 pF; 0.5 kΩ
Write/erase cycles	Semiconductor	100,000 — 500,000
Temperature cycling	MIL-STD 883/1010	-40 — 125 °C; 100 cycles / 0.5 h
Data retention	MIL-STD 883/1008	125 °C
Corrosion of contacts	MIL-STD 883/1009	24 h; 40 °C; 3 % NaCl

Each test has its own acceleration model. For the temperature humidity biased test (THB test) the acceleration factor AF can be determined according to the following model (see Equation (9.2)). The testing time (T_{test}) arises as a result from the division of the working time (T_{use}) and the acceleration factor (AF) (see Equation (9.1)).

$$T_{\text{test}} = \frac{T_{\text{use}}}{AF} \tag{9.1}$$

AF	=	acceleration factor
T_{test}	=	testing time
T_{use}	=	working time

$$AF = \left(\frac{Rh_{\text{stress}}}{Rh_{\text{use}}} \right)^3 * e^{\left[\frac{E_a}{k} * \left(\frac{1}{T_{\text{use}}} - \frac{1}{T_{\text{stress}}} \right) \right]} \tag{9.2}$$

Example values

AF	=	acceleration factor	
E_a	=	activation energy	(0.9 eV)
k	=	Boltzmann constant	($8.617*10^{-5}$ eV K^{-1})
Rh_{stress}	=	rel. humidity of the test climate in per cent	(93 %)
Rh_{use}	=	rel. humidity of application in per cent	(50 %)
T_{stress}	=	test temperature in K	(313.15 K)
T_{use}	=	application temperature in K	(298.15 K)

To calculate the acceleration factor AF for a temperature humidity test with 40 °C and 93 % relative humidity in reference to an application condition of 25 °C and 50 % relative humidity, the appropriate values must be inserted into Equation (9.2). For the example shown, an acceleration factor AF of 35 results for a testing period of 504 h which corresponds to an actual working time of 17,640 h or two years.

With this example it is not considered that the test is undertaken with active voltage. This leads to an additional extension of the tested simulated actual working time of more than the two years.

The lifetime of a cash card is three years; however this is usually not for its whole lifetime, but on average for a maximum of five minutes per day the card is in a smart card reader for the loading or unloading of cash. From this an actual working time for the three years of approximately 90 h results. The tested actual working time of 17,640 h is more than sufficient to cover the demand.

During the temperature humidity test, the corrosion behaviour of the chip is examined in relation to the particular behaviour of the contact pads of the chip. With each sealing compound of the chip module some humidity diffuses from the outside to the chip surface. Also at the interface of the sealing compound with the carrier tape humidity can reach the chip pad by the developing capillary effect of carrier tape roughness and then pass over the bond wires (see Figure 9.2).

In additional to humidity diffusion into the sealing compound, ion movement in the sealing compound and along the bond wires to the chip pads can be strengthened by the connection of an external current supply for the chip module.

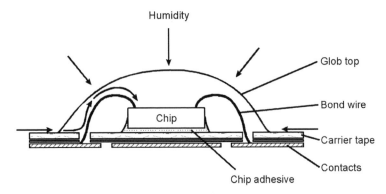

Figure 9.2 Possible ways for humidity to penetrate into a chip module

For example, if pure deionized water arrives at the aluminium pads of the chip, as shown in Equation (9.3), the following chemical reaction results:

$$Al + 3H_2O \Rightarrow Al(OH)_3 + 1.5H_2 \uparrow \tag{9.3}$$

This reaction forms a stable aluminium hydroxide layer, which actually prevents further water attack and slows the oxidation of aluminium down greatly. However, due to pollution in the sealing compound or on the chip surface from the wafer manufacturing process, ions are always available and deionized water is no longer present. Chlorine ions and sodium ions, in particular, can decompose the protecting aluminium hydroxide coat of the chip pads (see Equation (9.4)). The $Al(OH)_3$ is converted into the well-soluble $Al(OH)_2Cl$ compound for direct humidity attack.

Next chlorine aluminate hydrolyses with water to form aluminium hydroxide. The used up chlorine ions (Cl^-) generate hydrochloric acid (HCl).

$$Al + 2H_2O + Cl^- \Rightarrow Al(OH)_2Cl + H_2 + e^-$$
$$Al(OH)_2Cl + H_2O \Rightarrow Al(OH)_3 + HCl \tag{9.4}$$

The catalytic cyclic process again converts aluminium into aluminium hydroxide. This process continues for a considerable time until the water or the aluminium of the chip pads is completely used up.

Far-advanced corrosion can lead to electrical failure as the bond itself separates between the nail head and the chip pad.

Also, during the temperature humidity test the quality of the passivation layer of the chip can be tested. Through defects (holes, cracks, etc.) in the passivation layer the described ions can reach the aluminium conductor which is under the passivation layer and cause some corrosion. This can lead to a failure or functional fault of the chip.

A further important test is the lifetime test. Using this test, the lifetime of the chip with the user is simulated. The units are brought into a test chamber which is operated at test temperature of 85 °C. As in the case of the temperature humidity test, the chip module is electrically operated. If the lifetime of a cash card, which is three years, is to be tested then the test units are exposed for 21 days to these test conditions.

The live test has its own acceleration model, allowing the test time to be shortened.

A further specific test for the chip is data retention. With this test the memory cells, primarily the EEPROM cells, are tested as to how long they are able to hold the stored information. Usually data retention for 10 years is guaranteed by the semiconductor manufacturers. Obviously in order to check the data contents, the whole guaranteed 10-year duration cannot be supported. Again acceleration factors are used.

Using Equation (9.1) and Equation (9.5) the acceleration factor *AF* can be calculated. In order to achieve an acceleration, the chip is stored at test temperature of 125 °C in a test chamber for 21 days instead of at an ambient temperature of 25 °C. The 21 days result from a calculated acceleration factor *AF* of 353. Before start of the storage the EEPROM cells are written with particular data sample patterns.

$$AF = e^{\left[\frac{E_a}{k} * \left(\frac{1}{T_{use}} - \frac{1}{T_{stress}} \right) \right]}$$

(9.5)

Example values

AF	=	acceleration factor	
E_a	=	activation energy	(0.6 eV)
k	=	Boltzmann constant	($8.617*10^{-5}$ eV K^{-1})
T_{stress}	=	test temperature in K	(398.15 K)
T_{use}	=	application temperature in K	(298.15 K)

After the end of the testing period the cells should contain the same information as at the beginning of the test. Normally intermediate measurements are executed after one, seven and 14 days, in order to detect and analyse possible failures early.

The data retention test is important as the chip must not lose the information stored, for example the money value of a cash card, over its entire lifetime.

The temperature cycle test is best suited to test the total system chip, bonds and sealing compound. To enable this test, equipment with two test chambers is needed which are kept at different temperatures as shown in Figure 9.3.

During the test time the basket with the test units (chip module tape) is normally moved a hundred times from test chamber **I** (+125 °C) into test chamber **II** (-40 °C) and back. The test units remain for 30 min in every temperature zone. The surrounding test medium is air. After the temperature cycles, the electrical functionality of the test units is checked.

The temperature change causes mechanical bracing in the components of the chip module, due to their individual coefficients of expansion. The transition from the high-temperature test chamber **I** to the low-temperature test chamber **II** is particularly critical. Here weaknesses in the chip module like incorrect bonds, hair-cracks in the chip or weak points in the chip adhesive — which can lead to lift off of the chip from the carrier tape — are determined.

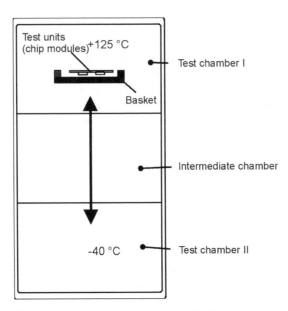

Figure 9.3 Schematic structure of test equipment for the temperature cycle test

The ESD (electrostatic discharge) test is designed to test the protective circuits within the chip in relation to ESD damage.

Ultraviolet (UV) radiation and the test in a magnetic field are intended to test the reliability of the individual components of the chip like those of the EEPROM cell in relation to UV radiation and magnetic fields. These two tests can lead to a loss or change of data content for insufficiently protected chips.

Lastly, we describe the write—erase test. With some hundred chips from different chip production lots the EEPROM cells are written and read again. With this test the reliability of the EEPROM cells and the control electronics is tested. If for example a cash card is put into a reader in order to execute a transaction, e.g. to load money, on some EEPROM cells write/readings are executed. The EEPROM should not fail, which could cause an error or can even lead to a total failure.

The test is frequently executed at 25 °C. To compare chips from different semiconductor manufacturers, the specified temperature is important with the specification of the value of the write—erase cycles. With higher test temperatures (for example, 85 °C) the value is lower, although it would be the same or higher if a temperature of 25 °C is assumed. A common value is 500,000 write—erase cycles at 25 °C.

We have presented an overview of some possible electrical reliability tests. Naturally, each chip manufacturer has additional in-house reliability tests in order to guarantee the quality of the final products.

9.3 Physical reliability tests

In addition to the electrical reliability tests the physical tests, also called the mechanical tests, are extremely important for the later reliability of the smart card.

In Table 9.6 a selection of physical tests are specified. In the second column the standard is listed which indicates the method with which the test is to be undertaken. The last column quotes the standards which define the parameters for the test. For the contact smart cards, tests from ISO 7816 1/2 are predominantly used. For the contactless card, in addition ISO/IEC 10536-1 is consulted. ISO/IEC 10373 describes the test conditions which test equipment is to be used, and the conditions in which the reliability tests must be executed.

Most tests must be undertaken with a test climate of 23 ± 3 °C and a relative humidity of 50 ± 10 %.

Table 9.6 Overview of selected physical tests for smart cards

Tests	Method	Parameter
Format	ISO/IEC 10373	ISO 7816-1
Surface profile of contacts	ISO/IEC 10373	ISO/IEC 7810
Dynamic bending stress	ISO/IEC 10373	ISO 7816-1
Dynamic torsion stress	ISO/IEC 10373	ISO 7816-1
Location of contacts	ISO/IEC 10373	ISO 7816-2
Card warpage	ISO/IEC 10373	ISO 7810 5.3.2
Embossing relief height of character	ISO/IEC 10373	ISO 7811-1
Delamination	ISO/IEC 10373	ISO 7810
Resistance to chemicals	ISO/IEC 10373	ISO 7810 514
Amplitude measurements	ISO/IEC 10373	ISO 7811-2 621622
Bending stiffness	ISO/IEC 10373	n/a
Flammability	ISO/IEC 10373	ISO 7810 512
Light transmittance	ISO/IEC 10373	n/a

The simplest test is measurement of the length, width and thickness of the smart card according to the values in Table 9.7.

Table 9.7 Dimension for a smart card without embossing

	Minimum (mm)	Maximum (mm)	Nominal (mm)
Length	85.47	82.60	85.72
Width	53.92	54.03	53.98
Thickness	0.68	0.84	0.76
Radius of the corners	2.88	3.48	3.18

The values from Table 9.7 are valid for smart cards without embossing. If a smart card with embossing is measured then the maximum permitted values are slightly increased.

In order to measure the position of the contact chip module in a smart card, the smart card has to be pressed (**F1**, **F2**) against three reference points **P1**, **P2** and **P3**, as shown in Figure 9.4.

During the measurement of the position of the contact area the smart card must be pressed with precise strength onto the measuring base plate, following this positions **A**, **B**, **C** and **D** are measured. The given positions (see Table 9.8) and tolerances of the individual contact areas (C1—C8) must be kept.

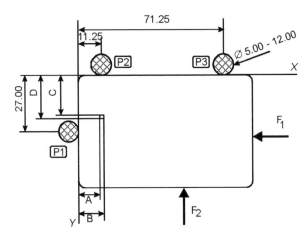

All dimensions in mm

Figure 9.4 Position of the contacts of a chip module on a smart card (here only one contact is shown)

Table 9.8 Position of the contacts C1 up to C8 relative to the upper left edge of the smart card

Contact	A (mm)	B (mm)	C (mm)	D (mm)
C1	10.25	12.25	19.23	20.93
C2	10.25	12.25	21.77	23.47
C3	10.25	12.25	24.31	26.01
C4	10.25	12.25	26.85	28.55
C5	17.87	19.87	19.23	20.93
C6	17.87	19.87	21.77	23.47
C7	17.87	19.87	24.31	26.01
C8	17.87	19.87	26.85	28.55

One of the most frequently required tests is the dynamic bend stress of the smart card. The smart card is inserted into a bending device as shown in Figure 9.5 and Figure 9.6.

Narrow and width side of the smart card

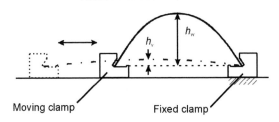

h_w = max. warpage

h_v = minimal allowed warpage in
the start position (2.00 mm)

Figure 9.5 Principle of ISO dynamic bend stress

The opposite clamp moves towards the fixed clamp and back again. First, the smart card is inserted with the contacts facing upward with long side into the bending device. With this movement the maximum deflection to h_w of the smart card is 2 cm. The smart card is bent at a rate of 30 bendings per minute. This occurs 250 times. Afterwards, the smart card is turned (contacts downward) and bent 250 times. Subsequently, the smart card is inserted with its short side into the bending device with the contacts upward and bent again 250 times. Here for the maximum deflection h_w is 1 cm. Next, the smart card is turned again and bent another 250 times. Thus, altogether a bending cycle of 1000 bends is completed. In between the changes after each 250 bends the smart card is tested for electrical functionality. After 1000 bends the smart card should not show any breaks in the card material, the chip module should still remain in the cavity and the chip should still be fully electrically functional.

Figure 9.6 Test machine for the dynamic bending stress test (source: Mühlbauer SCF 2300)

A further standard test is the dynamic torsion stress test. Figure 9.7 shows a bending machine for the torsion test. The smart card is clamped at the short side and twisted around $15 \pm 1°$. This test is executed with 30 twists per minute. After 1000 twists the smart card is examined for electrical functionality and also for possible breaks in the card body.

Using these dynamic bending stress tests the mechanical stability of the chip module is tested. In addition, the characteristics of the chip module in the smart card are tested. Also the chip module should not pop out of the cavity of the smart card.

For example, as a requirement for the health-insuring card, four times ISO bends are required, this means 4000 bending cycles.

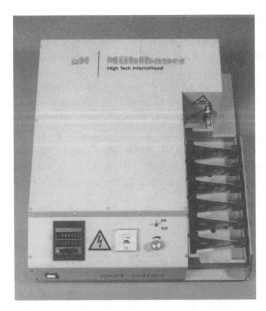

Figure 9.7 Test machine for the dynamic torsion stress test (source: Mühlbauer SCT 2400)

For some time efforts have been made to simulate dynamic bending on the basis of the finite element method (FEM) using a computer. The smart card is divided into a number of sections (elements) of finite size (see Figure 9.8). Following this, the sections are idealized, for example, by triangles. After definition of the load boundary conditions (direction of force, amount of force) an FEM program can plot the load on the smart card.

It is clear that simulation using a computer can be very useful. However, because of the different materials and connection techniques which are used in the smart card, at the end of the development of a new smart card, only the real mechanical load can show how stable the smart card actually is.

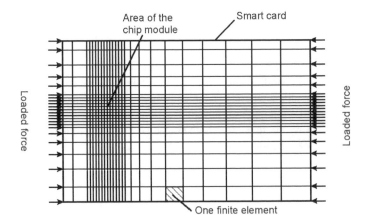

Figure 9.8 Idealized smart card subdivided into finite elements

As a further requirement, the smart card must not set toxic materials free in daily use. Also, it must be resistant against certain chemicals, for example 5 % of salt-bearing solution, 50 % ethyl glycol and gasoline following ISO 1817. The valuation criterion is that the smart card must always remain electrically functional. If a card contains a magnetic stripe it too must be still functional.

An important test for the laminated smart card is examination of adhesion between the plastic foils themselves and adhesion of an overlay foil on top of a colour print.

Using a sharp knife a 25 mm broad strip, as shown in Figure 9.9, is cut in the card. As the next step, part of the upper foil is taken off with a sharp knife, as shown in Figure 9.10, upward 6 mm. Afterwards, a strongly responsive adhesive strip is stuck to the splice straps. The test is carried out in such a way that the card and adhesive are clamped into a mounting plate and pulled apart. The strength required is compared with a standard.

There are still more reliability tests which can be looked up in the relevant standards.

As already mentioned, smart card manufacturers and many smart card customers have their own reliability tests established which are not specified in the standards.

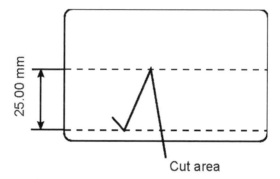

Figure 9.9 Cut area for the delamination test

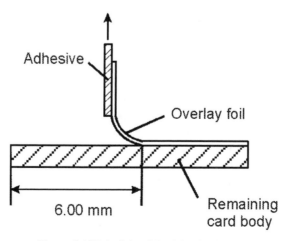

Figure 9.10 Principle of the delamination test

For example, there is a so-called hand test to A and B. Here the smart card is adapted in such a way that from the centre point of the chip module the smart card is cut away to a distance of 15 mm. If the test card is cut out of the long side of the smart card, it is called hand test A; from the short side it is called hand test B (see Figure 9.11).

Afterwards the adapted smart card is pushed together as in Figure 9.12 and Figure 9.13 for 1.5 mm or 3.0 mm.

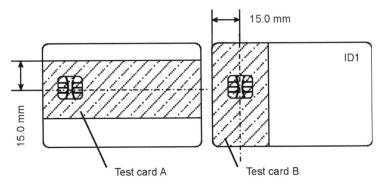

Figure 9.11 How to cut out a test card for the hand tests A and B out of an ID1-smart card

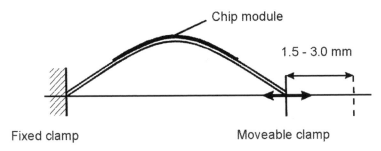

Figure 9.12 Warpage of the cut out test card for the hand tests A and B

The smart card is inserted once into the clamp where the chip module is arranged upward, and once downward. The failure criteria is the adhesive strength of the chip module in the smart card and electrical functionality. If the test card is pushed together by 3.0 mm, normally only the adhesive strength of the chip module in the smart card is tested. At 1.5 mm the chip module should also remain electrically functional. This test makes particularly high demands of the connection technique and stability of the chip module as the load axis runs directly through the chip module and affects high mechanical forces on the chip module and its bonding with the smart card.

This chapter overviewed different reliability tests. Beyond this there are different testing methods for the remaining card items, for example for the holograms, magnetic stripes and signature panels.

Figure 9.13 Test machine for hand tests A and B, B is shown (source: TESA)

After introducing various card bodies, chip modules, connection techniques, production techniques and the most used reliability tests in the preceding chapters, now the last production step of the smart card, — smart card personalization — will be described.

10 Personalization and mailing of smart cards

While the telephone card is not bound to a single user and can be used by different people without restriction, most other smart cards are intended for use by an individual only (or a certain group of people). The manufacturing process which makes the unique link between a person and the smart card reference is so-called personalization.

Reference to the person can also be made indirectly. Smart cards for mobile phones (GSM smart cards) are referenced to a user identification number with the personalization. After the conclusion of a contract the user identification number is assigned to the user.

The smart card can also be linked to a device or a machine instead of a person. In this case the smart card functions like a key. Everyone, who is in possession of this card, has access to the respective device or machine.

For personalization, some or all of the following specifications can be characteristic of the smart card:

- Information for the card system
- Particulars of the card owner
- Identification number
- Signature
- Photo
- Fingerprint

The above information can be visible on the smart card and, in addition, be coded into the chip or on the magnetic stripe. Beyond this there is a set of safety-relevant data which can be coded into the chip, for example:

- PIN (personal identification number)
- Cryptographic keys (partial or whole)
- Further biometric data

The following features can be brought in for automatic recognition:

- UV coding
- One or two dimensional bar coding

With some card systems it is additionally necessary that certain data for the card system are produced with the personalization. This generates a reference for the respective card to the card system. Depending upon the safety relevance of these files they must be encoded or secured in a different way and transferred to the central computer of the system.

It can be further necessary to have a reference between the smart card and user and make certain data accessible for the card owner in a safe way. Such data is for example, the PIN number. The PIN is sent to the user in a closed security envelope. The secured PIN letter is arranged in such a way so that it cannot be opened unseen. Externally, it is provided with camouflage printing which consists of an amassment of many different digits (the so-called digit salad). This printing prevents the PIN from being read in strong transmitted light. In order to exclude abuse of the smart card by unauthorized users, it is necessary to never allow the PIN and smart card to be made available at the same time. For optimal security for the card owner use of the card is only allowed in the system if authentic acknowledgement from the smart card owner that he/she has received the card and intact PIN letter is obtained.

10.1 Work flow for personalization

Figure 10.1 shows a generalized flow for personalization. Depending upon the specific request of the card system deviations from this flow are conceivable. The flow shown in the figure serves to understand the systematics, which can be made valid for different configured operational sequences. The steps in the flow of the personalization process are numbered **1, 2, 3**... The steps which are numbered **1.1, 2.1, 2.2, 2.3**, etc. give out data or materials. The sequence of the numbering in the flow does not show the temporal consequence of the exchange.

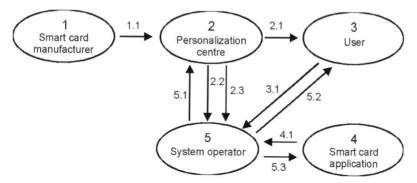

Figure 10.1 Flow for smart card personalization

1.1	Smart cards from card manufacturer to the personalization centre
2.1	PIN letter from personalization place to the user
2.2	Personalized cards from the personalization centre to the system operator
2.3	System data from the personalization centre to the system operator
3.1	Data of the card owner to the system operator
4.1	Registration of the smart card application with the system operator
5.1	Personalization data of the system operator with the personalization centre
5.2	Delivery of the personalized smart card to the user
5.3	Authorizing of the card application by the system operator

In the flow shown, the personalization place **2** receives the cards **1.1** from the card manufacturer **1** and personalization data **5.1** from the system operator **5**. The personalization centre carries out the card personalization and creates PIN-letters, which are sent directly to the card owner. Depending upon grouping of the personalization, system-relevant files **2.3** can result with the personalization. These must be transferred from the personalization place to the system operator. The personalized cards are sent from the system operator to the card owner. The card owner using the card and PIN is able to access services which are offered through authorized service providers. The service provider is authorized by the system operator and receives the necessary keys for specified communication with the personalized smart card. For communication between the service provider and the smart card special terminals are used. These are personalized in nearly the same way as the smart card through the system operator.

10.2 Personalization technologies

There are different methods for smart card personalization, which differ in size, of data, inscription methods of the cards and procedure for data communication. Some typical examples are listed in the following:

Credit cards

Most credit cards are embossed. The embossed information contains the surname and the first name of the owner, the date of issue and date of expire of the card and a card identification number. The card is signed personally by the card owner on the signature panel. The card identification number is additionally imprinted in the signature panel by a thermal transfer embossing procedure. During personalization of the credit card the magnetic stripe is also coded. Using the stored information on the magnetic stripe receipts can be created automatically. There are also credit cards containing a chip which has to be coded during the personalization.

In addition, some credit cards have a picture of the card owner on the rear side.

Eurocheque cards

Most Eurocheque cards are personalized using a laser personalization procedure. The readable personalization contains the first name and surname of the card owner, and the account and card number. Also, the last four digits of the account number are written on the hologram by laser engraving in order to provide additional security against falsification of the card. The magnetic stripe is also personalized. The personal secret PIN number is coded on the magnetic stripe. Some Eurocheque cards carry covered machine-readable features, the so-called MM. With the personalization, a code related to MM is likewise coded on the magnetic stripe.

Figure 10.2 shows a Eurocheque card with a microprocessor.

One of the functions of the microprocessor in the Eurocheque card is to act as an electronic purse. The electronic purse allows the owner to transfer funds from his/her bank account into the card at special bank terminals. The owner can also pay very small amounts with the Eurocheque card at public telephones and at various machines e.g. cigarette machines, without using his/her PIN.

Figure 10.2 Example of a personalized Eurocheque smart card with a microprocessor (source: Giesecke & Devrient)

GSM cards

For the GSM digital mobile telephone system the entire intelligence for the authentification within the network is installed in the SIM card (subscriber identity module) and is separated from the mobile phone itself. The advantage of this separation is that both the SIM card and mobile phone can be developed separately from each other. While telephone devices have been equipped with constantly new comforts and additional general functions and miniaturized at the same time, the SIM card has always been developed towards higher security and higher non-standard functionality (as for example, a personal directory).

With the personalization of the GSM card the safety functions are coded into the chip, whereby the personalization data refers to a user identification number which is only assigned to the user after the signing of a contract.

The GSM card is also personalized with an additional card identification number. The inscription usually takes place via laser engraving.

ID cards

Identity cards are used in the areas of employee identity, patient and health information and driving licences.

ID cards contain both non-standard (different for each card) and common security characteristics (the same with each group of cards). They can be humane (without additional electronic devices managed by a control person) security characteristics, as well as machine-readable characteristics.

ID cards usually carry personal particulars — the photo and signature of the owner. Additionally they can contain a fingermark and other biometric features of the owner. Using the laser-engraving procedure it is possible to engrave all these features onto the card in one processing step. Therefore it is necessary to integrate information from manual input data, record the photo from digitized medium or a video, and accommodate the signature and fingermark from a digitized medium in a data record, which is then engraved on the card by lasers.

The safety requirements of ID cards can be very different. For example, the patient insuring cards in Germany do not need high safety requirement as the information stored (name, assurance number) in the chip is not confidential. Very high safety requirements against falsification are required for identity cards to prevent unauthorized modification of the information on a card or total falsification by unauthorized production of identification cards.

Figure 10.3 shows an ID card with various card items.

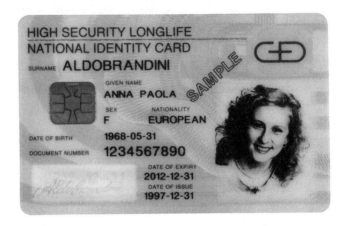

Figure 10.3 Sample ID card with a laser-engraved picture (source: Giesecke & Devrient)

The MLI (multiple laser image, also see Section 2.4.8) is especially suitable as a security feature for ID cards due to its high abrasion firmness and hence higher lifespan and also good recognizability in weak light. As the MLI is engraved with a laser, it optimally can be integrated in laser personalization machines.

Figure 10.4 shows a modular structured personalization machine which can be configured depending upon the required features.

Figure 10.4 Personalization machine (source: Giesecke & Devrient)

The feed module is a revolving table with 10 stations. Appropriate card input and card output modules are arranged around the feed module and depending upon the configuration different handling modules, for example a card check module (optical check of the inscription with a video inspection unit), chip coding module, magnetic stripe coding module, laser engraving module, etc., are present.

Decentralized personalization

The previously described procedure of central card personalization and subsequent shipping of the card is not desirable in some cases. With some applications, for example employee identity cards, membership cards, customer documents of identification, etc., it is desirable that the card can be personalized directly and locally. The expenditure of time for personalization can be larger than the cycle time of a personalization machine. Very often the security requirements for these cards are not as high as with bank cards or identity cards. For such applications, peripheral decentralized personalization is an optimal solution. Also by decentralized coding of smart cards, a secure functionality of the chip can be established.

A suitable procedure for decentralized personalization is thermocolour printing. Using this procedure a digitized picture and inscribing can be brought on the card surface. For this, a heated needle transfers the colour from a transfer foil onto the surface of the card body. By use of special coloured hot embossing foils and an embossing raster in the basic colours (thermal sublimation procedures) even fastidiously coloured pictures can be transferred to the card surface. Thermotransfer printers are available which are additionally equipped with a chip coding station with which smart cards can be personalized visually and electrically. If for this procedure a non-white blank card with safety printing and additional safety elements is used, increased security can be manufactured against falsification of the inscription and photo.

This is due to the fact that the card carries security printing or UV printing on the surface, which functions as background for the thermal transfer printing. Falsification attempts on the thermal transfer printing lead to damage of the background and can be determined with a check.

A further procedure for decentralized personalization is mechanical engraving. The card surface is partially covered with a printed colour area in a colour which is in contrast to the colour of the underlying card surface. During personalization, a fine burin engraves individual points on this printed surface, as a result the printed surface is penetrated in the respective point and the background shows up.

Ink-jet printers are also available for card personalization. These are, however, mostly used for printing a serial number on the card, during production.

10.3 Smart card mailing

The security of a system can be compared to a chain which has to carry a load. It is sufficient that only one segment within the chain is designed too weakly for the load which can be carried, and the chain tears. The strength of the other segments of the chain is completely irrelevant.

Smart card mailing is such a segment in the security chain — it is not only a question of logistics.

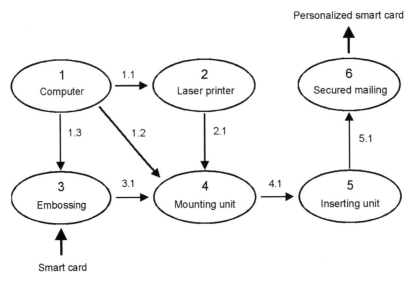

Figure 10.5 Flow of personalization and mailing of the personalized cards

1.1	Data for address printing for the letter
1.2	Data for aligning card/letter
1.3	Data for embossing of the card and coding the magnetic stripe
2.1	Printing letter
3.1	Personalized card
4.1	Letter with fixed personalized cards on the letter
5.1	Letter with card and attachments in the envelope

Figure 10.5 shows the flow of personalization and mailing of credit cards.

Using remote transmission or transportable data carriers, the central computer **1** checks in the received data and converts it into the required formats and transfers them to the handling stations.

The laser printer **2** personalizes the pre-printed letter.

The embossing system **3** personalizes the cards, and codes the magnetic stripe and chip.

The card mounting system **4** adds the personalized card to the letter.

The inserting system **5** inserts the letter into the envelope with additions like advertising or manuals.

The safety mailing sends the letters directly to the credit institute or through the postal system to the user.

11 New applications for smart cards

In addition to the representation of different applications for standard smart cards, their production possibilities and much more besides, here we point out further application types and versions for the smart card. In future besides the classical contact areas of the usual standard smart cards there will be new input/output elements established e.g. displays, keyboards and different sensors. Also the shape and areas of application will change. This chapter gives a brief overview of the future.

11.1 Smart cards with electronical input/output devices

At present the smart card industry is looking at ways to equip new smart cards with extended functions. There are some prototypes with new input and output elements. The most promising input/output items belong to displays and keyboards.

Displays may be used to show for example, the contents of the electronic purse, which is stored in the chip, to the smart card owner.

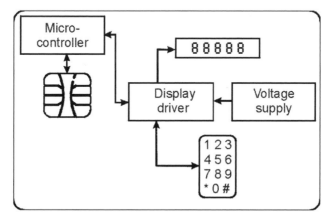

Figure 11.1 Schematic arrangement of the individual components for a smart card with a display, keyboard and voltage supply

A keyboard on the smart card can be used as an input medium for the PIN (personal identification number). Hereby the smart card would be a closed system in itself. This becomes important for future applications from the point of view of safety, for example when paying by transfer for goods on the Internet. Figure 11.1 shows the interaction in principle of the individual components.

For these new elements to be used in large numbers the displays used must become thinner and more flexible, and have to be more economical. Also the power input of the displays must be reduced to the necessary minimum.

In order that the displays can also function as autonomous systems, smart cards with a display and keyboard require their own voltage supply. For voltage supply, brought in batteries or solar cells can be integrated into the card body. Some development activities have started in both these areas. Figure 11.2 shows an application for a smart card with an integrated display and switch.

Figure 11.2 Example smart card with a display, integrated battery and switch (source: Giesecke & Devrient)

The necessary production techniques still have to be developed for application of these new card items in the smart card.

11.2 Smart cards with sensors

Besides input and output elements, smart cards have been developed which contain integrated sensors. Sensors for temperature, moisture measurement and capacity measurements are currently in the development stage. In particular, sensors which support biometric identification procedures, will have widespread future applications for the smart card as they extend the system security.

The biometric identification procedures are based on the identification of unmistakable physical characteristics.

There is a variety of procedures already, which use the following biometric features: the individual signature of a person, whereby the dynamics of the signature are carefully analysed. Then there is the possibility of analysing the keystroke on a keyboard whereby the typical writing rhythm is detected. Also, a face carries characteristic features which can be analysed through a digitized picture. Moreover, speaking an agreed word can be analysed over the frequency spectrum and assigned to a particular person.

All these individual features of a person could be detected using sensors; nevertheless, it will be very difficult to integrate them into the smart card.

There is, however, a further feature which is unique for each person and which can be detected by sensors integrated in the smart card: the fingerprint. The pattern of finger ridges is detected by capacitive sensor arrays having the size of a finger. Afterwards the data are analysed using biometric algorithms. Hence, it is possible to make a secured authentification for a person. This already occurs with an error recognition rate of 1:1,000,000.

Figure 11.3 represents the structure in principle of a capacitive sensor for finger structure recognition. The sensor method is based on the different capacity values between ridges lines and the sensor which have a lower value than the value between valleys and the sensor. This due to the fact that there is air acting as dielectric in the valleys.

Presently there are two manufacturers developing a finger sensor: Firstly, we have a narrow sensor array from ST Microelectronics company, with which the finger must be pulled slowly over the sensor field.

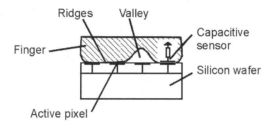

Figure 11.3 Principle structure of a capacitive sensor for fingerprint recognition

Secondly, from the Infineon company comes the so-called FingerTip[TM] sensor. This has an active sensor surface of 11.1 * 14.3 mm² with a resolution of 224 * 288 pixels. With this finger sensor, the finger can be placed directly on the sensor field. Figure 11.4 shows the FingerTip[TM] sensor for application with computer keyboards.

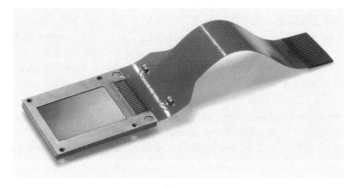

Figure 11.4 FingerTip[TM] sensor (source: Infineon)

The fingerprint recognition sensor suffers from the same problem as the displays. It also requires a 5 V voltage supply. Both elements have to be very thin in order to accommodate the FingerTip™ sensor and battery in one smart card. Also, these components will cause additional problems in maintaining the usual flexibility of the smart card after integration. All these problems will most likely be solved allowing mass production in a few years.

11.3 New applications for smart cards

The smart card in its current form as a ID1-smart card will change appreciably. In fact, the plug-in SIM (ID-000) for GSM application in mobile telephones is a modification of the smart card (ID1). Thus, the plug-in SIM (ID-000) is very similar to the smart card. The most recent products on the smart card market, e.g. the MultiMedia Card™ from the Infineon company (see Figure 11.5), already deviate considerably from the standard smart card.

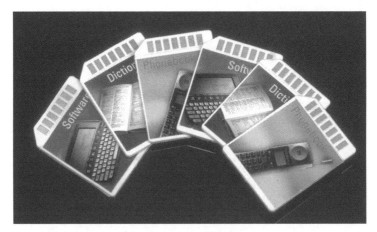

Figure 11.5 MultiMedia Card™ (source: Infineon)

Starting with the external dimensions of 32 * 24 * 1.4 mm³, the available memory also differs from conventional smart card chips which have 64 KB as opposited to the substantial 2,4 or 8 MB memory of a MultiMedia Card™. A further distinction is the position, arrangement and number of electrical contacts. The MultiMedia Card™ has seven electrical contacts, arranged differently to the contact position for the standard smart card, and the electrical contacts, as shown in Figure 11.5, are in one line at the edge of the card.

The MultiMedia Card™ is used as a mass storage device for advanced applications such as PDAs (personal digital assistants), hand-held PCs, palmtops and audio applications. The advantage over conventional storage media, e.g. floppy diskettes, lies in the fact that with the MultiMedia Card™ no mechanical parts have to be moved. Also the data access times of under 3 µs are faster than floppy diskettes.

Additionally to the entertainment electronics, the MultiMedia Card™ is used in telecommunications as a mobile directory or as a travel and hotel guidebook, which may be plugged into a mobile telephone. In future more applications will develop for the MultiMedia Card™.

Also the storage capacity of the MultiMedia Card™ will continue to increase — up to 128 MB for the year 2001.

In addition to the MultiMedia Card™, the future will generate further applications for the smart card; Nevertheless, the standard smart card will remain with us for a long time due to its convenience in use and flexibility.

.

12 Short cuts

ABS	Acryl butadiene styrene
ACA	Anisotropic conductive adhesive
ACF	Anisotropic conductive film
AF	Accelerator factor
AFNOR	Association Française de Normalisation (French Standards)
Al	Aluminium
AQL	Acceptance quality level
As	Arsenic
ASK	Amplitude shift keying
ATM	Automated teller machine
ATR	Answer to reset
Au	Gold
B	Boron
b.c.c.	Body-centred cubic
Bit	Binary digit
C4-Prozess	Controlled collapse chip connection
CD-ROM	Compact disc read-only memory
CLK	Clock
CMOS	Complementary metal-oxide semiconductor
CMS	Chip management system
COB	Chip on board
COT	Chip on tape
CPU	Central processing unit
DBG	Dicing before grinding
DES	Data encryption standard
DIN	German Industrial Standard ‚Deutsche Industrienorm'
DIP	Dual in-line package
DNA	Deoxyribonucleic acid
DRAM	Dynamic random-access memory
Ec	Eurocheque
EEPROM	Electrically erasable programmable read-only memory
EPROM	Erasable programmable read-only memory
ESD	Electrostatic discharge
eV	Electron volt

F	Frequency
f.c.c.	Face-centred cubic
FEM	Finite element method
FERAM	Ferro electrical random-access memory
FR4	Epoxy resin, glass-fibre reinforced
Ga	Gallium
Ge	Germanium
GND	Ground
GSM	Global System for Mobile Communication
HBM	Human body model
I/O	Input/output
IC	Integrated circuits
ID	Identification
IEC	International Electrotechnical Commission
In	Indium
ISO	International Organization for Standardization
JEDEC	Joint Electron Device Engineering Council
LTPD	Lot tolerance per cent defective
MIL-STD	Military Standard
MLI	Multiple laser image
MMC	MultiMedia Card
MOS	Metal-oxide semiconductor
MOSAIC	Microchip on surface and in card
Nc	Not connected
NCP	Non-conductive paste
OVI	Optically variable ink
P	Phosphorus
PC	Personal computer
PC	Polycarbonate
PDA	Personal digital assistant
PET	Polyethylene terephthalate
PETG	Polyethylene terephthalate glycol-modified
PIN	Personal identification number
PROM	Programmable read-only memory
PVC	Polyvinyl chloride
RAM	Random-access memory
ROM	Read-only memory
RST	Reset

SAM	Secure application module
Sb	Antimony
SCQL	Structured card query language
SEM	Scanning electron microscope
Si	Silicon
SIM	Subscriber identity module
SMT	Surface mounting technology
SRAM	Static random-access memory
STARCOS	Smart card chip operating system
T	Transmission protocol
TAB	Tape automated bonding
THB	Temperature humidity bias
UBM	Under bump metallization
UV	Ultraviolet
V_{cc}	Operating voltage

13 Bibliography

Brockhaus, Naturwissenschaften und Technik Sonderausgabe. Mannheim: Brockhaus, Band 1–5.

Card Forum. Ausgabe 12/97.

Card Forum, November 1998, 5. Jahrgang.

Card Forum, Ausgabe 11/98.

Einführung in die Klebetechnik. Loctite Deutschland GmbH. 2. Auflage, 1991.

Finkenzeller, Klaus: RFID-Handbuch – Grundlagen und praktische Anwendungen induktiver Funkanlagen, Transponder und kontaktloser Chipkarten. München, Wien: Hanser, 1998.

Hoppe, Bernhard: Mikroelektronik. Herstellungsprozesse für integrierte Schaltungen. Würzburg: Vogel, 1998.

Hoppe, Bernhard: Mikroelektronik. Prinzipien, Bauelemente und Werkstoffe der Siliziumtechnologie. Würzburg: Vogel, 1997.

Klein, Martin: Einführung in die DIN Normen. Stuttgart, Berlin: Beuth, Köln: Teubner, 1980, 8. Auflage.

Kunz, Land, Wierer: Neue Konstruktionsmöglichkeiten mit Kunststoffen. WEKA Fachverlag für Führungskräfte. Stand: November 1998.

Lau, John H.: Flip Chip Technologies. New York: McGraw-Hill, 1995.

Markt und Technik. Nr. 23 vom 4.6.99.

Rankl Wolfgang, Effing Wolfgang: Handbuch der Chipkarte. Aufbau – Funktionsweise – Einsatz von Smart Cards. München, Wien: Hanser, 1999.

Renesse, Rudolf L. van: Optical Document Security. 2nd Ed. London: TNO Institute of Applied Physics, 1997.

Sautter: Leiterplatten mit oberflächenmontierten Bauelementen. Würzburg: Vogel, 1988.

Smart Card Technologies and Applications. November 16—18, 1998.

Systeme (Elektronik-Magazin für Chip-, Board- & System-Design); Heft 2, Februar 99.

Trockenes Entfetten und Aktivieren mit Niederdruckplasma. Kunststoffberater, Ausgabe 10/93.

Van Zant, Peter: Microchip Fabrication: A Practical Guide to Semiconductor Processing. New York: McGraw-Hill, 1997.

14 Appendix

14.1 Company overview

The following table provides a short overview of different smart card manufacturers and companies which deliver material and machines for the production of smart cards.

Company	Short description
3M www.mmm.com	• Adhesive systems for smart cards
AmaTech GmbH www.amatech.de	• Contactless smart cards • RF-ID keys • Contactless pre-laminated inlays
American Express www.americanexpress.com	• Credit institute
Atmel www.atmel.com	• Microcontrollers • Secure memories • Contactless chips
Austria Card www.austriacard.at	• Smart card manufacturer
Boewe Systec AG www.boewe-systec.de	• Dispatching machines for plastic cards
Bürkle Robert GmbH www.buerkle-gmbh.de	• Laminating presses
Cardel Ltd Pixmore Avenue Letchworth. Herts. GB SG6 1JG	• Heat-activatable adhesives
Cards & More www.cardsnmore.com	• Smart card manufacturer • Card printing systems

Company	Short description
Cards&Devices Neue Str. 67 D-99846 Seebach	• Smart card manufacturer
Cardxx www.cardxx.com	• Smart card manufacturer
Cicorel GmbH www.cicorel.ch	• Chip module carrier tape (PCB)
ComCard www.comcard.de	• Smart card manufacturer
CUBIC www.cubic.com	• Contactless readers • Contactless system 'Go Card'
Cubit Electronics Wilhelm-Wolf-Str. 6 D-99099 Erfurt	• Contactless smart cards
Dage www.dage-group.com	• Bond tester • X-ray • SAM
Dai Nippon Printing Co., LTD www.dnp.co.ip	• Smart card manufacturer
Datacard GmbH www.datacard.com	• Card personalization machines • Identification systems
Datacolor www.datacolor.de	• Smart card manufacturer
De La Rue Card Systems www.delarue.com	• Smart card manufacturer
DELO Industrial Adhesives Landsberg www.delo.de	• Conductive silver pastes • Glob tops • Underfiller
Dexter GmbH www.dexelec.com	• Conductive silver pastes • Glob tops • Underfiller
Digicard GmbH www.digicard.at	• Smart card manufacturer
Diners Club www.dinersclub.com	• Credit institute
Disko www.disko.com	• Wafer-processing machines

Company	Short description
DuPont www.dupont.com	• Conductive silver pastes
Eastman Chemicals AG www.eastman.com	• Plastic foils
EM Microelectronics Marin www.emmarin.com	• Smart card manufacturer • Contactless inlays
Europay International www.europay.com	• European organization for electronic payment systems (EUROCARD, MasterCard, Maestro, Cirrus, Eurocheque)
FCI MCTS 37, rue des Closeaux F-78200 Mates la Jolie	• Chip module carrier tape (PCB)
Flextronic www.flextronics.com	• Chip module carrier tape (PCB) • Contactless antennas
Freudenberg Mektec www.freudenber.com/mektec/	• Contactless antennas
Gemplus www.gemplus.fr	• Smart card manufacturer
Giesecke & Devrient GmbH www.gdm.de	• Smart card manufacturer • Personalization machines • ATMs
Heidelberg www.heidelberg.com	• Digital printing machines
Heraeus www.heraeus.de	• Chip module carrier tape
Hitachi www.hitachi-eu.com	• Microcontrollers • Secure memories • Contactless chips
Hologram Industries Edewechter Landstr. 28 D-26131 Oldenburg	• Holograms
ID Data Systems Ltd, Wansell Road Weldon North, Corby NN17 5LX, GB	• Smart card manufacturer
Indigo www.indigonet.com	• Printing machines

Company	Short description
Infineon www.infineon.com	• Microcontrollers • Secure memories • Contactless chips • MMC
ISO www.iso.ch	• International Standards Organization
Kaba Holding AG www.kaba.com	• Door automation • Access control • LEGIC contactless system
Klöckner Plast Werk Gendorf D-84504 Burgkirchen	• Plastic foil manufacturer
KSW Microtec www.ksw-microtec.de	• Smart card manufacturer
Leonhard Kurz GmbH & Co. www.kurz.de	• Hot-stamping machines • Magnetic foils • OVD foils
Loctite www.loctite-europe.com	• Glob tops • Underfiller
Louda Systems www.louda.de	• Punching machines • Gathering machines • Magnetic tape layer • Hot-stamping machines
MasterCard International www.mastercard.com	• Credit institute
Melzer Maschinenbau GmbH www.melzer-germany.com	• Smart card equipment producer
Mühlbauer AG www.muehlbauer.de	• Smart card equipment producer
Multitape www.multitape.de	• Chip module carrier tape (PCB)
NedCard Lagelandseweg 48 N-6545 CE Nijmegen	• Chip module carrier tape (PCB)
Novacard Informationssysteme GmbH www.novacard.de	• Smart card manufacturer

Company	Short description
Oberthur Smart Cards www.oberthurusa.com	• Smart card manufacturer
Orga Kartensysteme www.orga.com	• Smart card manufacturer
Otto Künnecke GmbH www.kuennecke.com	• Mail processing systems
PacTech www.pactech.de	• Wafer bumping service • Flip-chip test kits
Panacol-Elosol GmbH www.panacol.de	• Underfiller • Glob tops • Dispensing units
PAV Card GmbH www.pav.de	• Smart card manufacturer
Philips www.semiconductors.philips.com	• Microcontrollers • Secure memories • Contactless chips • Readers
Polytec GmbH www.polytec.com	• Underfiller • Glob tops
PPC Card Systems www.ppc-card.de	• Smart card manufacturer
Printoplast AG www.printoplast.com	• Smart card manufacturer
Protechno Card www.protechno-card.com	• Personalization machines • Mailing machines • Thermal transfer printers
Ramtron International Corporation www.ramtron.com	• Microcontrollers • Contactless chips • FRAM technology
Rimec GmbH www.rimec.de	• Offset printing machines • Single printing machines • Counting machines
Ruhlamat www.ruhlamat.de	• Smart cards production machines
Schlumberger www.slb.com	• Smart card manufacturer

Company	Short description
Sempac www.sempac.com	• Smart card production machines
Shellcase www.shellcase.co.il	• Shellpack
SICPA www.sicpa.com	• Security inks
Sokimat Zone industrielle CH-1614 Granges (Veveyse)	• Antenna suppliers
ST Microelectronics www.st.com	• Microcontrollers • Secure memories • Contactless chips • Readers
Sun Microsystems www.sun.com	• Java platform
Tesa www.tesa.de	• Heat-activatable adhesives
Texas Instruments www.ti.com	• Microcontrollers • Secure memories • Contactless chips
Trüb Switzerland www.trueb.ch	• Smart card manufacturer
VISA International www.visa.com	• Credit institute
Wilden Kunststoff- und Gerätetechnik GmbH www.wilden.com	• Moulding machines
Winter AG www.winter-ag.com	• Smart card manufacturer

14.2 Overview of actual chips for smart cards

Overview of actual chips used for smart cards from the major silicon manufacturers.

14.2.1 Memory chip

Table 14.1 Memory chip from Infineon

Product	ROM	PROM	EEPROM	Remark
SLE4406/06E	16 bit	56 bit	32 bit	
SLE4436E	16 bit	185 bit	36 bit	
SLE5536S	16 bit	185 bit	221 bit	
SLE4404	16 bit	144 bit	256 bit	
SLE4466	16 bit	226 bit	272 bit	
SLE4432/42	—	32 bit	256 bit	
SLE4418/28	—	1024 bit	1024 bit	

Table 14.2 Memory chip from ST Microelectronics

Product	ROM	PROM	EEPROM	Remark
M14C64	—	—	64 bit	
M14128	—	—	128 bit	
M14C256	—	—	256 bit	
ST1200	—	—	256 bit	
ST1305B	—	—	192 bit	
ST1331	—	—	272 bit	
ST 1355	—	—	272 bit	

14.2.2 Contactless memory chip

Table 14.3 Contactless memory chip from Infineon

Product	ROM	PROM	EEPROM	Remark
SLE44R35	—	—	1 KB	Mifare® (Philips)
SLE55R01	—	—	128 byte	
SLE55R16	—	—	1024 byte	

Table 14.4 Contactless memory chip from ST Microelectronics

Product	ROM	PROM	EEPROM	Remark
Mifare Standard MF1S50	—	—	1 KB	
Mifare Light MF1L50	—	—	384 Kbit	

14.2.3 Microcontroller chips

Table 14.5 Microcontroller from Infineon

Product	ROM	RAM	EEPROM	Remark
SLE11C01S/U	3.5 KB	1280 B	16 KB	
SLE44C80S/U	17 KB	256 B	8 KB	
SLE66C20S	31.5 KB	1280 B	2 KB	0.5 Kbyte for RMS
SLE66C80S	31.5 KB	1280 B	8 KB	0.5 Kbyte for RMS
SLE66C160S/U/P	31.5 KB	1280 B	16 KB	0.5 Kbyte for RMS
SLE66C320S/U	31.5 KB	1280 B	32 KB	0.5 Kbyte for RMS
SLE66CX160S	31.5 KB	1280 B+ 700 B	16 KB	0.5 Kbyte for RMS; crypto engine
SLE66C640P	134 KB	4 KB	64 KB	
SLE66CX640P/PU	135 KB	5 KB	64 KB	1100 bit crypto engine 64 bit DES accelerator

Table 14.6 Microcontroller from Hitachi

Product	ROM	RAM	EEPROM	Remark
H8/3112	24 KB	1312 B	8 KB	576 bit crypto engine
H8/3114	32 KB	2560 B	16 KB	1024 bit crypto engine
H8/3153	32 KB	1 KB	16 KB	
H8/3158	46 KB	1 KB	16 KB	
H8/3166	32 KB	1 KB	2 KB	
H8/3164	48 KB	3 KB	32 KB	

Table 14.7 Microcontroller from Philips

Product	ROM	RAM	EEPROM	Remark
P8WE6008	8 KB	768 B	32 KB	
P8WE6032	32 KB	1280 B	32 KB	
P8WE5033	32 KB	2300 B	96 KB	Crypto engine
P8WE5017	16 KB	2304 B	64 KB	Crypto engine

Table 14.8 Microcontroller from ST Microelectronics

Product	ROM	RAM	EEPROM	Remark
ST16600	6 KB	—	512 B	
ST16601	6 KB	—	1 KB	
ST16SF42	16 KB	—	2 KB	
ST16SF48	16 KB	—	8 KB	
ST16SF4F	16 KB	—	16 KB	
ST19SF08	32 KB	—	8 KB	
ST19SF16	32 KB	—	16 KB	
ST19SF32	32 KB	—	32 KB	
ST19CF68	23 KB	—	8 KB	Crypto engine

14.2.4 Microcontroller chips — dual-interface

Table 14.9 Contactless microcontroller from Infineon

Product	ROM	RAM	EEPROM	Remark
SLE66CL160S/U	31.5 KB	1280 B	16 KB	100 %; 10 % ASK

Table 14.10 Contactless microcontroller from Philips

Product	ROM	RAM	EEPROM	Remark
P8RF5008	48 KB	2304 B	8 KB	100 % ASK
P8RF5016	64 KB	2304 B	16 KB	100 % ASK
P8RF6004	32 KB	1208 B	4 KB	100 % ASK; crypto engine
P8RF6008	48 KB	1208 B	8 KB	100 % ASK; crypto engine

Table 14.11 Contactless microcontroller from ST Microelectronics

Product	ROM	RAM	EEPROM	Remark
ST16R820	8 Kbyte	—	518 byte	ISO 14443-2 Type B
ST16RF52	22 Kbyte	—	2 KB	ISO 14443-2 Type B
ST16RF58	22 Kbyte	—	8 KB	ISO 14443-2 Type B

Index